THESE KIDS ARE DRIVING ME CRAZY

The EMPOWER Method to Simultaneously Keep Your Sanity and Your Students in Your Classroom

Kristen Miller

Cover Design: Kristen Miller

Editing: Jessica Mulholland, Erika Parsons

Praise for Kristen Miller's
THESE KIDS ARE DRIVING ME CRAZY

"I just finished Miller's book and I am so excited to share it with my staff. Miller has been able to consolidate some of the most impactful techniques for managing challenging behaviors in one, easy to ready resource. Miller's experience as both a teacher and an administrator provides a balanced perspective. Her collection of stories of how she has managed different situations in her career using the EMPOWER method spoke to me and will to others who read it. I recommend this book to anyone looking for a "one-stop shop" for techniques that will manage challenging behaviors."

– Donna Marie Cozine, Ed.D., CEO
Renaissance Academy Charter School of the Arts

"I may have a unique viewpoint while reading 'These Kids Are Driving Me Crazy" because I read this book as a student (not an educator) who is both immensely concerned with the current educational system in the USA and also greatly inspired by those who are actively tackling the structural issues embedded into our system. I am thankful and grateful to know educators like Kristen Miller and many others who take steps to employ caring about their students wellbeing and understanding that the best way to help students thrive is to understand them as a human and meet them where they are to inspire them and motivate them to become the best learner they can be. Not only does Kristen do this in her classroom, but she has gone the extra mile by seeking out different pedagogical strategies, executing them, AND sharing them in order to reform education. Our education system doesn't need a light tap, it needs a wrecking ball to be rebuilt and the strategies and EMPOWER method in this book aims to do just that. If you are looking into becoming an educator, need a revitalized view of teaching and your career path, or are enthralled by the good souls who have taken on working to better our education system, I can't recommend it enough."

–Jessie Vera, Former Student

"Kristen provides a well thought out blueprint for success with data to back it up. Her real life stories provide insight and credibility to the research she presents. I appreciate her willingness to be vulnerable and lead by example in all of the strategies outlined in her book. I am excited to embrace and practice the EMPOWER Method! This is a must read for teachers and administrators!"

–Michelle Guardino, High School Teacher

"In this book Miller perfectly blends anecdotal stories and advice. She guides readers through how to identify the underlying causes of students' frustrating behaviors and methods to positivity turn it around. A must read for teachers (new and seasoned), for school site administrators, and for anyone with a vested interest in helping kids reach their full potential."

–Jaime Bonato, High School Teacher

"This book is a must read. I appreciated the honesty in how hard teaching can be. This was definitely written from a "with" perspective when looking at the social discipline window. Honest reflection and moving forward with intentionality resulted in success for all. This book aligns well with Restorative Practices and PBIS as well as social-emotional learning. Easy read with helpful background knowledge and tips to implement."

–Amanda Foster, Teacher & District-Level Staff

"To educators everywhere, especially those just entering the the profession. I would encourage you to read Kristen Millers book, These kids are driving me crazy. It will add to your knowledge a practical, heartfelt and encouraging way to be the teacher you imagine yourself to be. Teachers, are in the caring industry. This book will allow you to use that part of yourself, to reach children, who will know their teacher cares. Kristen outlines her journey into teaching the "tough students", the ones who didn't care, who slept through class, who disrupt for the sake of not understanding the material, and brings you into the real world of caring enough to ask, how are you doing? Kristen has walked in your shoes, she knows the reality of the profession. These kids are driving me crazy is, worth the read and using the practical tools to inspire others to do what they love, teach the next generation of learners about being life long learners."

–Terri Walsh, Instructional Assistant

Dedication

To Sophie and Nic:
my heart, my home, my happiness.
You inspire me to be the best version
of myself every day.

Table of Contents

Chapter 1: You're Not Alone

*"Kids cussing out teachers is commonplace
these days."*
— Dr. Bruce Perry

The teaching profession is crazy nowadays. Absolutely crazy. If you're not sure whether you have the stamina to stay in this field for the next forty years, you're not alone. According to *The Wall Street Journal*, "Teachers and other public education employees, such as community-college faculty, school psychologists and janitors, are quitting their jobs at the fastest rate on record, government data shows."

What is that rate? Research from PDK International shows that 50 percent of teachers leave the profession in the first five years of teaching. It's no wonder – in addition to the long hours and low pay, teachers face some of the most outrageous student behaviors ever and receive such little respect from the public that after a few years, they decide it's just not worth it. So, what does teaching actually look like nowadays?

- Kids not listening during lessons.
- Kids sleeping during class.
- Kids blatantly disrespecting their teachers.
- Kids cussing teachers out.
- Kids rifling through teachers' personal belongings.
- Kids engaging in fist fights in classrooms.
- Kids reading at a fourth-grade level, despite being a senior in high school.
- Kids operating at a third-grade math level, despite being a senior in high school.
- Kids stealing from other kids.
- Kids stealing from teachers.
- Kids bringing weapons to school.
- Kids bringing drugs to school.
- Kids bringing alcohol to school.
- Kids taking apart small pencil sharpeners so they can take the blade to the bathroom and cut themselves.

- Kids ganging up on teachers and pushing their buttons to try and get them fired.
- Kids having sex on campus.
- Kids coming to school drunk, high, or both.
- Kids getting virtually no consequences for cussing a teacher out.
- Kids punching in school windows.
- Kids being sexually assaulted on campus.
- Kids being murdered as a result of gang involvement.
- Teachers losing their sanity because of out of control behavior.
- Teachers being asked to change grades so students can graduate or play football.
- Teachers being told, "It's your word against the student's."
- Teachers being yelled at by parents.

- Teachers losing jobs because of lawsuits.
- Teachers losing jobs because of political corruptness in the system.
- Administrators being forced to tolerate outlandish behavior because they aren't backed by the school board.
- Educators being fired because of site-level affairs.
- Educators being fired for doing the right thing.
- Educators being fired for speaking up about illegal or unethical processes, procedures, or occurrences.
- Race wars.
- Cyber-bullying.

The words "hot mess" come to mind, and that's exactly what it is ... a big, hot mess. We've allowed our education system to be shaped by powerful politicians, special interest groups, and lawyers — all at the expense of our students' and staffs' well-being.

If you can relate to any of the above scenarios, know that you're not alone. Every situation listed above is something I personally encountered and worked through during my thirteen-year career in the public education system. I spent most of those years as a high school math teacher. I also had short assignments as an Advancement Via Individual Determination (AVID) teacher, Career Technical Education (CTE) teacher, vice principal, and an intervention coach. I've seen it all.

With all this mess, it's no wonder people struggle to stay in this profession. Generally, people who enter the teaching profession do so because at their core, *they want to help kids*. They want to empower kids and inspire them to be their best. Then all the things in the list above start to wear them down year after year, and for some, week after week. I've seen several teachers start teaching and after five or six weeks, they quit. It's complete madness!

If you're reading this book, you likely are struggling with the profession and are trying to decide about your future. You love teaching. You love working with the kids, but some days it just feels like too much. It's overwhelming. You may be experiencing anxiety, depression, and burnout, and you're worried that if things don't change, you may have to start from scratch in a new profession.

But you still have hope. You don't want to leave this profession because you realize just how important it is. You have hope that somewhere out there, someone recognizes (or a group of people recognize) the reality in our schools and they're working to make the shifts necessary to empower teachers to empower students. Guess what? There are. And I am one of them.

It took me thirteen years to find the root of the problem, and I'm so excited to share these strategies with you. If you have the courage and bravery to make some changes in your classroom, and use the tools and strategies outlined in this book, you'll be able to *teach* your kids *and* maintain your sanity. You'll also be able to enjoy teaching while making a massive difference in your students' lives and your own life.

Take a minute and think about a truly amazing moment in your teaching practice. Maybe it's one student, maybe it's an entire class, maybe it's just a realization you came to that allowed you to make the difference you were born to make in education. Let that feeling really sink in.

That is why you are here. That's why you're reading this book. That's why you know deep down you want to be in this job for the long haul.

In this book, I'm going to ask you to take some risks and really identify the kind of teacher you want to be. It's going to feel scary at times, but don't worry — I'm right here with you. Together, we can move mountains *one student at a time*. The more mountains we move, the more of a massive shift in education we can make together.

So, let's get started.

Chapter 2: I've Been There, Too

My first year of teaching was one giant learning curve. I stepped into the classroom and was given my own set of 150 students to work with — but I had no former or formal teaching experience. At that point, I had taken a few teaching credential classes and spent four to five hours after my normal engineering day job going to school to learn how to become a teacher. So I'd taken maybe four or five classes, and was plopped into a high-poverty high school teaching mostly Pre-Algebra (the equivalent of seventh grade math) to primarily juniors and seniors. And these students were in this class not because they weren't smart — they straight up didn't care.

Why didn't they care? This question was *always* in the back of my mind as I traipsed through my thirteen years of different teaching environments, different colleagues, and different demographics — all with one commonality. But let's not get ahead of ourselves. To really understand how I figured this out, you need some context.

As I mentioned, my first teaching assignment was at a high-poverty school (mostly students in poverty, but with a small population of affluent students). As with every school year, the first week or two began with kids on their absolute best behavior. Then they settled in. I have very vivid memories of specific interactions (they are flying through my mind as I type) that illustrate exactly what was going on in my classroom.

The funny thing is, had an administrator walked into my classroom during what is my first memory of confronting a disengaged student, it seemed like all was in order. Kids were facing forward and taking notes. It was quiet and everyone appeared to be paying attention. I looked like a rock star teacher.

The problem was, this one student — who initially had his head up, paper on his desk with a few words from the lesson sloppily written — started closing his eyes as if he were falling asleep. I kept an eye on him as I continued with the lesson.

I waited until he was full-blown asleep to call him out: "Open your eyes, son." Honestly, at the time, I don't even know where that came from. I've never used the term "son" since. My uncle was a principal for many years and often used that term when addressing male students, so it just sort of came out. The student's eyes opened quickly and he realized I wasn't going to let him get away with sleeping in class. I was setting a precedent. You come to Miss Miller's class to *learn*. That's it. The end. Anything other than learning would not be tolerated.

This changed the game. Many of these students (and teachers as I would come to find out) stopped caring and therefore had incredibly low expectations of themselves or their students. So, stepping into a classroom where learning was the expectation was culture shock.

After the "honeymoon period" wore off, kids started to figure out how to piss me off, and they did it *repeatedly*. They understood the expectation, and while many went along with the expectation of learning, there were still plenty in each class to test me. They discovered that if they said or did something that went against my expectation, I got frustrated. When I got frustrated, my face turned bright red and I would have a very reactionary emotional response. It was a fun game for them. They figured out my buttons, my triggers, and took great enjoyment pressing each one of them every opportunity they could. As I would later learn, there were many things they were dealing with in their lives to cause them to behave this way, but that explanation didn't help me right there in the moment. I was drowning.

By September of my first teaching year, I felt done. I was mentally done, physically done, and emotionally done. One of my favorite images that depicts teachers' overall demeanors from beginning of the school year until the end is shown below. Only problem is that this wasn't the *end* of the school year, it was a month in during my first year of teaching *ever*. I knew if I wanted to survive in this profession, I'd have to make some massive changes. So, I did.

Teacher at the BEGINNING of the School Year

Teacher at the END of the School Year

After a particularly frustrating day when a fist fight almost broke out in my class (and kids could see I was flustered and didn't know how to handle it), one of my very challenging students very casually mentioned, "Aww, it's her first fight. She doesn't know what to do." I was bright red and on the verge of tears. I had no idea what to do, but I was *pissed* that one of my students used my frustration for her own entertainment. I went home and did an internet search for "classroom management high school." *Many* websites came up talking about how to:

- Make sure you have a warm-up on the board.
- Provide consistent structure and routine.
- Utilize effective academic strategies.
- Scaffold instruction.

Stated simply, there was a lot of very unhelpful information. But I decided to give one book a chance: *Teaching with Love and Logic*. I ordered it immediately with rush shipping so I could finally ease some of the daily frustrations I felt. As soon as it came, I read the first few chapters and implemented the strategies right away. It transformed my classroom!

Don't get me wrong. Kids were still misbehaving. But thanks to *Love and Logic*, I learned how to manage my classroom without being as affected by daily ridiculous, outlandish student behavior. This was truly a game-changer.

Implementing these strategies allowed me to get a better handle on my students' behavior. The instances of disruption, defiance, and disrespect dramatically decreased and we were actually *learning*! I managed to get every student to take notes in class. I got all the kids attempting to do their classwork *and* their homework. I was thrilled!

Then they had their tests. There were so many Fs. I was upset and confused, but also intrigued to figure out exactly what was going on. I began to talk individually with students who failed and tried to really get to know them as people. I started with, "What happened on this test?" I'd get a variety of answers ranging from, "I dunno," to "My dad hit me last night so I couldn't study." Some of these kids' stories and experiences were mind blowing and truly devastating.

I was hooked. I made a habit of scanning the room every class period. When students had a sad or *"just not themselves"* looks on their faces, I made a point of asking them two questions:

- "Are you doing okay?"
- "What's going on?"

This has been the single-most effective "strategy" I've utilized in my practice all thirteen years. I learned *so much* about my students just by asking these two questions. If they opened up, I'd continue asking questions — and often would hear about the massive amounts of trauma these kids were subject to when they went home each day. Some of the situations I've helped students through over the years include:

- Physical abuse at home.
- Sexual abuse at home.
- Emotional abuse at home.
- Neglect at home.
- Absent parents.
- No parents or foster care.
- Gender identity issues.
- Sexuality issues.
- Rape.
- Sexual assault.
- Child trafficking.

- Murder.
- Gang involvement and activity.
- Drug and alcohol abuse.

Some of the things these kids deal with are heartbreaking, yet we rarely talk about them. Whenever it came to professional development or training or student success, it was all about academics. As an educator, I often heard questions and statements like:

- "How many of your students are proficient on the standardized test, Ms. Miller?"
- "How many of your students are passing your class, Ms. Miller?"
- "What teaching strategies are you using, Ms. Miller?
- "I think you could have scaffolded that math lesson a little better."
- "How are you differentiating instruction for English learners and students with IEPs?"

- "Today in our staff meeting, we're going to talk about Webb's Depth of Knowledge and Bloom's Taxonomy. We really need to up the rigor of our instruction."

It was absolutely maddening. I dedicated my entire teaching career to serving my students, not *just* academically, but as human beings. I cared about them as *people*, not just as test scores or statistics for the California School Dashboard. I often had this idea in the back of my head:

HOW CAN WE EXPECT KIDS TO LEARN RIGOROUS
CONCEPTS LIKE APPLYING LOGARITHMS
TO REAL-WORLD SITUATIONS,
WHEN THEY'RE
STRUGGLING
TO JUST
SURVIVE
FROM
DAY-
TO-
DAY
?

I made it my mission to check in with as many students as possible each day and ultimately provide a safe space for them. They didn't need any more trauma at school; they were experiencing too much at home already. Those of you who have worked with high-poverty populations know how incredibly trying it can be. After three very fulfilling and passionate years at my first school, I was spent. I needed to recharge. A position opened at a very high-performing charter school right down the road, so I applied.

It broke my heart to move on from the first school of my career. But I realized it was taking a huge toll on me. I needed a calmer environment. Working at the high-performing charter school was a dream. There were virtually no behavior issues. I think I wrote one referral in the entire five years I was there (whereas at my first teaching job, I wrote one referral each class period every day). At my new school, the kids were engaged, they were on-task 99 percent of the time, and they *all* cared about their grades. It wasn't uncommon for me to have twenty-five As in a class of thirty students. It was a completely different environment than where I was before. I could finally breathe.

I was given a challenge my second year at this school to double the proficiency on the Algebra 2 STAR test. At the time, I was teaching almost all the sections of Algebra 2. I figured it would be a piece of cake, given the environment I was teaching in. Compared to my former students, most of these kids had zero issues to worry about outside of school. Most had all the guidance, encouragement, and support they needed. I implemented a few new programs to increase student accountability and was on my way. I waited anxiously for the test scores to come out the following school year to see if I met my goal.

I did. I more than doubled the proficiency on the Algebra 2 STAR test. But more than doubling the proficiency still only put us at 43 percent proficiency. I was confused. I implemented everything I could think of. I taught every concept covered on the test, I put the correct accountability systems in place, yet we were still only at 43 percent?! What the heck?!

I took a closer look at my students from that point on. I noticed I had very few students who had been exposed to trauma of any kind, but these were still teenagers going through the typical changes of adolescence. They were still learning about themselves and their identities, and we didn't give any of that attention at school. It was all about the academics; all about the rigor; all about test scores and data. It wasn't about them *as people*.

I started to really miss my first school. My passion was, and still is, working with underserved and underrepresented populations. I came to the realization that I missed this so much and I needed to get back out to a high-poverty school again. A position opened at a high-poverty school closer to home. I jumped at the chance. Even though there was an emphasis on test scores and data at this school, there was at least the acknowledgement that these are kids who have many other things going on. I could acknowledge and support them in more ways than just academically.

It was a great few years. I took on additional leadership positions, created and implemented a Science Technology Engineering Art Math (STEAM) pathway to add to the school's already impressive CTE portfolio. I was in my element and truly thought I found my "forever school" and my "forever career path." Then it all changed when a student got injured in my classroom. More about this later.

I made an abrupt career change and went into administration at a high-performing school in a different district. I fell in love with the job. I once again felt like I found my "forever school" and my "forever career path," but life had different plans for me. I had no idea what kind of hornets' nest I was walking into; the year I was there, laws were broken, education codes were violated, board policies were violated, and the office politics were off the charts. This year forced me to decide where I wanted my career to go. It turns out it was in neither education administration, nor any one specific school.

I started an organization to help students and teachers; to inspire them, empower them, and ultimately help them be the best they could be. I wanted to make massive shifts in education. I saw what I believed to be a very big solution to a very big problem in education, and I couldn't wait to partner with a school to test my theory out. I was incredibly fortunate to have landed at a school part-time teaching math, part-time under my organization with the hopes that I could implement my ideas and see my mission and vision transform into reality. And I did. It was an amazing year.

I did some informal action research to test my theory that we need to spend less time teaching academics and more time teaching people skills (also known as social-emotional skills). We need to spend more time building relationships with our students and parents. By doing this, we will see our students reach higher levels of achievement than ever before. I am still incredibly grateful I landed in a school with a new administration whose vision aligned with mine. I created and implemented programs in my own classroom (explicit social-emotional lessons in addition to math), as well as schoolwide in conjunction with the site leadership team (mostly administrators and counselors). I collected data throughout these programs to evaluate the effectiveness of my "out of the box" programs and efforts.

The results were amazing. My students grew more academically than other math students on campus. Students who got into multiple fights the previous year and had a huge number of discipline infractions now had zero fights and very few discipline infractions. The lessons I learned during this school year were truly life-changing, ground-breaking, and confirmed my prior assumptions about what kids needed from their schooling.

I learned that we've been missing the mark for years; our obsession with academic focus, academic strategies and test scores will never bring up test scores like district leaders want. Why? Because society has shifted so dramatically in the last fifty years that the needs of students (and society) are no longer the same. Kids don't *need* teachers to teach them strictly academic content anymore. The increase in technology has caused a decrease in humanity. Kids (and not just kids, people in general) don't know how to be human anymore, and it's getting in the way of their individual success and the ability to achieve their fullest potential.

We can do better. I *have* done better. And I am so excited to share the detailed tools and strategies I used to help students achieve their highest potential and staff enjoy teaching. It all starts with behavior management. Under the radical behavior you're seeing in your classroom lies a lot of deeply unresolved issues that are seeking a safe space to get resolved.

Chapter 3: How This Works — The EMPOWER Method

The ultimate crux of my success in education was that learning who my students were as people, finding their strengths, helping manage their challenges, and ultimately empowering them to be their best was the best thing I could do for them. One of the ways I could manage all of this was working with school staff and empowering them with the right tools and strategies to be successful. If the adults on campus feel empowered to make change, that energy will transfer to the students both directly and indirectly.

This book is designed to give you a set of tools to empower you to handle all the unsaid, unseen, and unspoken challenges buried deep in the K-12 education system. Oftentimes, we have challenges we encounter directly in our classrooms, and this book will give you direct strategies to manage those as well, but we often don't realize how much what is going on in your classroom is a spider-like web of complex, intricate, entangled, and often dysfunctional systems rooted in politics and appearance. We're going to address these issues because this is our reality. If we don't begin having *real* conversations about *real* issues, we'll never see *real* change. And our education system is desperate for *real* *change.*

The tools in this book can be read out of order, but I encourage you to start at the beginning, as the beginning chapters will give you immediate strategies to be utilized in your classroom today. I call this system The EMPOWER Method for education. You can use it to effectively manage and mitigate specific issues in your classroom, as well as issues outside your classroom directly affecting your ability to effectively do your job. The steps of The EMPOWER Method are described as follows:

E–Eliminate Bad Behavior

In chapter 4, you will be given a variety of tools to help eliminate bad behavior immediately.

These tools include utilizing the most current trends in education: strategies from the *Love and Logic* book, Positive Behavioral Interventions and Supports (PBIS), Restorative Practices (RP), Trauma-Informed Practices (TIP), and Social-Emotional Learning (SEL). As soon as you're done reading this chapter, you can implement the specific tools to immediately shift your teaching practice.

M–Maintain Positive Administrative Relationships

In chapter 5, we will cover what the vice principal job looks like from the teacher's perspective. We'll learn what a day in the life of a vice principal looks like, including teacher support and school politics. We'll also cover steps to help teachers work effectively with administrators and provide a tool for administrators to effectively work with teachers.

P–Push Through the Politics

In chapter 6, we will cover a variety of situations that are highly politically charged and relate directly to what's going on in your classroom, as well as tools to maintain dignity and grace when facing political pressures.

O–Own It

Chapter 7 is devoted to self-exploration; we'll look at our own behavior management style and implicit bias, we'll identify our own triggers, typical responses to these triggers, and then create constructive responses to these triggers.

W–Why Not Try Something New?

In chapter 8, new strategies are outlined to incorporate into your classroom, including: community-building circles, mindful breathing, mindful walking, mindful listening, wellness writing, daily greetings, taking time to learn your students' names, celebrating your students, snowball fights, sharing your own stories, and being vulnerable.

E–Empower Yourself

Chapter 9 is devoted to self-care. We discuss the importance of acknowledging struggles of the profession, knowing your triggers, using worksheets or academic films when you're not "feeling it," using sick days, walking it out during heated and escalating classroom interactions, and challenging ourselves to take better care of ourselves.

R–Relationship-Build

The final step is all about building relationships. Chapter 10 talks about the importance of equity when it comes to discipline and consequences, and specific ways to build relationships with students while towing the line. It explains the importance of being vulnerable and how to use your own life stories in your instruction. It also discusses how practicing these methods will not make your students walk all over you. In fact, it will do quite the opposite; it will make them respect you and work harder for you than they ever have before.

The EMPOWER method is detailed and informative, yet incredibly simple and easy to implement. It is a method backed by data and one that will truly transform your classroom if implemented with fidelity. I'm not going to lie and say each step is easy to implement. The idea of implementing each of these steps is incredibly simple, but having the courage and the bravery to take the risk needed to implement these practices in your classroom is not for the faint of heart.

What do you have to lose? Together we can move mountains by EMPOWERing ourselves to be our best and meet our own needs so we can better meet the needs of our students. It all begins with real talk.

I believe in you.

Let's do this!

Chapter 4: Eliminate Bad Behavior

"If it's predictable, it's preventable."
— *Gordon Graham*

If you're anything like me, before you became a teacher you envisioned a classroom with twenty-five perfectly behaved children on the edge of their seats listening to each and every word you said. When you were teaching the lesson on solving equations for x that you intentionally created with love, every student was enthralled with what an awesome teacher you were and each was stoked to come to your class every day.

Then you actually got your own classroom.

The first year I taught, I was at a high-poverty high school in Sacramento, California. It was my dream job. I was teaching "Transitions to Algebra 1" to predominantly juniors and seniors. If you're not familiar with this class, it's essentially sixth and seventh grade math, but you're teaching it to seventeen- and eighteen-year-olds. These kids were checked out. They didn't seem to care about much in life, let alone the concepts I was teaching in my class. They gave me a run for my money.

The first week or so started out okay; they followed along with my lessons fairly-well. But shortly after we got into a routine, their true behavior came out. Kids tried to sleep in class, would blatantly talk over me when I was trying to get through a lesson, and generally exhibited a huge amount of disrespect.

But I was so passionate about what I was doing. I truly believed I was going to be like Michelle Pfeiffer in *Dangerous Minds*, where she turned these unmotivated, gangster kids into high-performing scholars. I distinctly remember a situation where a student named Trey kept quietly making fun of a female student who sat next to him. I asked Trey to stop, and he would say, "Okay, okay," but he continued making fun of her under his breath while I was writing math concepts on the board. Eventually she got so frustrated with him that she yelled at him to stop, which is when I intervened. I turned around and looked at the female student, at which point she said, "He's still sayin' stuff!" So, I walked over to Trey, got extremely close to him and said, "What's going on, dude? What are you saying?" He adamantly denied he was saying anything at all, and his continuous denial of his actions got under my skin. So, I got closer to him and said loudly and sternly in his face, "You want to say something? Okay, say something, say something to me." He got quiet, quickly. I learned a lot from that interaction, as well as many others in my first few years of teaching, all of which launched me into a thirteen-year-long informal action research project about behavior management.

Right around September of my first year of teaching, I was drowning. Kids weren't paying attention in class. They didn't care and I cared way too much. In a typical day, kids threw things across the classroom, made snide comments to each other to attempt to start fist fights, passed drugs hidden in ball point pens back and forth in class, and were generally distracted, disruptive, and defiant while they were supposed to be learning. I had only been employed as a teacher for a month, but I was fried. I desperately searched the internet for "classroom management high school students." That's when I came across the book, *Teaching with Love and Logic*. After that, I was on my way. I only ended up reading the first five or six chapters, because what I learned in those chapters transformed my classroom immediately. I'm not going to lie and tell you it eliminated *all* bad behavior in my classroom, but it drastically reduced the incidents of bad behavior from that point on.

In this chapter, I am going to share with you many of the tools and strategies I used over my thirteen years in education, most of which I was doing by trial and error. I continued doing the things that worked and stopped doing the things that didn't. I have come to learn that all of the tools and strategies I've used fall into one of several categories that now have explicit frameworks for teachers to utilize in their classrooms:

- *Teaching with Love and Logic*
- Positive Behavioral Interventions and Supports (PBIS)
- Restorative Practices (RP)
- Trauma-Informed Practices (TIP)
- Social-Emotional Learning (SEL)

Love and Logic strategies, the ones that transformed my classroom, are all about choices, respect and accountability. I started teaching without any formal training, and managing my classroom was trial by fire most days. Kids would misbehave. I would get frustrated and angry, which turned my face red and caused my whole body to tense up. Students loved it. It was a show for them, and certainly more interesting than learning how to solve equations for x. So anytime they could do or say something to get under my skin, they would, and my emotional roller coaster would ignite. I would get off task, and they would then no longer have to learn the lesson. I was drowning. They were clearly in control and sinking my classroom. The graph below, courtesy of Dr.

Bruce Perry of the Child Trauma Institute, is a perfect depiction of the types of interactions I was having on the daily. The kids were reactive, I was overwhelmed, and together we were co-dysregulated.

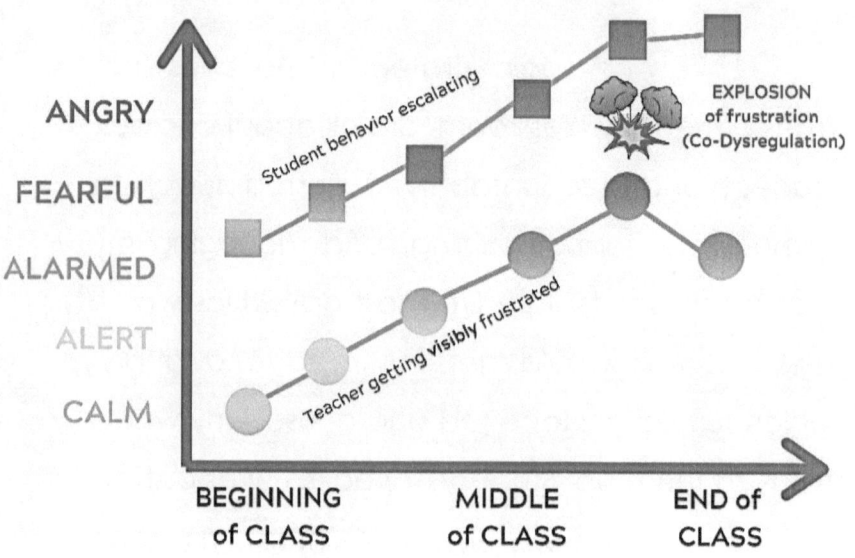

So, when I learned about *Love and Logic*, I implemented the strategies immediately. This is essentially how it works:

- Student says or does something to make me (or another student) mad.

- I calmly respond to the student misbehaving and say to him/her, "Okay Johnny, here are your choices. You can continue misbehaving and receive a class suspension, or you can stop and we can continue on with no more problems."

- At this point, the student tries to argue with me, saying that the other student, "was doing things to make me mad. It wasn't my fault. You're picking on me."

- I ignore all the responses from the student and repeat, "You can continue misbehaving and receive a class suspension, or you can stop."

- Naturally, the student continues to try to argue, so I simply say, "Okay, go ahead and head to the vice principal's office." I then walk over to the phone, call the office and tell the secretary, "Johnny is on his way to the office. He is suspended from class and should be there in the next minute or two."

- Johnny continues trying to argue, but as he begins to see that I'm not riding the emotional roller coaster as I typically do, it's no longer fun for him and he leaves the classroom.

This may seem overly simple and redundant, but there are two key parts to this process that make it effective: *choices* and *flat affect*. Choices are key in life. Take a minute and think about a time where someone just made a choice for you without your say. How did it feel? I'm guessing not great. As human beings, we thrive in situations where we feel like we have some control of our life, and that control often comes in the form of various choices. Think of something as simple as dinner. If someone told you every day what you were having for dinner and you couldn't choose, you'd probably go mad after a while. Same goes here. Now you may be saying to yourself, "But you're not really giving them a choice, because ultimately you suspended the kid anyway." However, even though it might *appear* that way, the kid got a choice before the suspension was ever issued. I very calmly say to him/her, "You can continue doing what you're doing and receive a class suspension, or you can stop, and we can continue on with no more problems." Seems small and miniscule, but *highly* important. By giving your student a choice, you are also showing them respect, which is *invaluable* when it comes to classroom discipline. The table below shows common infractions and how you can change your language to reflect choice:

INFRACTION	WHAT YOU SAY	YOUR TONE/ BODY LANGUAGE	FOLLOW-UP *(If student doesn't change behavior)*
Student throwing things in the classroom.	*"Your choice is to stop throwing things or receive a detention."*	**FLAT** Do **not** engage in power struggle or show emotion.	Assign Detention or Other Consequence*
Student talking during the lesson.	*"Your choice is to stop talking or continue talking. If you continue talking, I'll have to assign you a detention."*		Assign Detention or Other Consequence*

Student talking back to you.	*"Your choice is A or B."*	**FLAT** Do **not** engage in power struggle or show emotion.	Repeat choices **two times, then no more discussion.** Student has elevated to the next level by blatantly ignoring your requests.
Student blatantly ignoring your choices or requests (aka defiance)	*"Go ahead and head to the office. I will fill out your suspension paperwork shortly."*		Send student to the office and call home.

Tip: *Get in the habit of assigning lunchtime detentions for accountability purposes; allow them time to go get their lunch from the cafeteria (10-15 minute leeway), then return directly to your classroom. If they don't show up, they automatically go in the "Student blatantly ignoring your choices or requests (aka defiance) category in the chart above.*

The second crucial part of this interaction is the flat affect, also known as, "not going on the emotional roller coaster" with the kid. Many kids who aren't focused in class are struggling with deep issues that we can't even contemplate. Their lack of focus isn't because they are "bad kids," which is often the label they're given by teachers and other adults in their lives. They have likely had many struggles before they landed in your classroom and are just trying to survive day to day. They are used to chaotic environments, and chaos is their "normal." So, if things get too calm in your classroom, they feel uncomfortable and need to act out to maintain their level of normal.

Obviously, this doesn't help you manage your classroom, but the part that *will* help is keeping your cool. You may want to engage with kids in the ridiculous arguments they give you for why they shouldn't be getting in trouble, but *don't do it. It's a power struggle, and if you engage and get visibly frustrated, it will give them a sign that they're getting to you, and they'll continue because it becomes a game that they want to win.*

If you keep your cool and continue repeating the phrase, "Your choice is _____ or _____," without getting frustrated, there will no longer be a game to play and you will be in control. I recommend you only verbally give the choice two times, and on the third time you give the student the choice, you say the words out loud, "Your choice is _____ or _____. I'm going to walk away now, and your action will tell me what my next action will be."

Again, no power struggle. You are in control. Eventually you'll want to swing back around and talk with Johnny in a private setting to see what's going on with him, but for now, you just need to be able to teach your lesson, and if he's disrupting it, he needs to go.

The last and most important part of *Love and Logic* to acknowledge is accountability. Oftentimes teachers throw out threats to students like, "If you don't stop, I'm going to suspend you from class." But then they don't actually suspend them from class. All this does is shows your students that they're in control, which will make your life more miserable in the long run, because as soon as they realize they can do or say anything and won't get in trouble for it, that's exactly what will happen. This is such a common trap for teachers! It makes sense, though. Most of us don't like confrontation, and when you hold students accountable, you are confronting them at the highest level. But you have to stay strong! When you give your student a choice for their poor behavior, one of which is a detention or class suspension, you must follow through and actually assign the detention or class suspension; if you don't, you're setting yourself up for a very long and challenging year. Stay strong and stay accountable. You got this.

Positive Behavioral Interventions and Supports (PBIS)

PBIS is a big buzz word these days. Many schools and districts are scrambling to infuse PBIS practices to help better support their students. So, what is PBIS anyway? We already know what it stands for, "positive behavioral interventions and supports," but what exactly does that mean? PBIS has three tiers, and the following is a description of each tier and what it does.

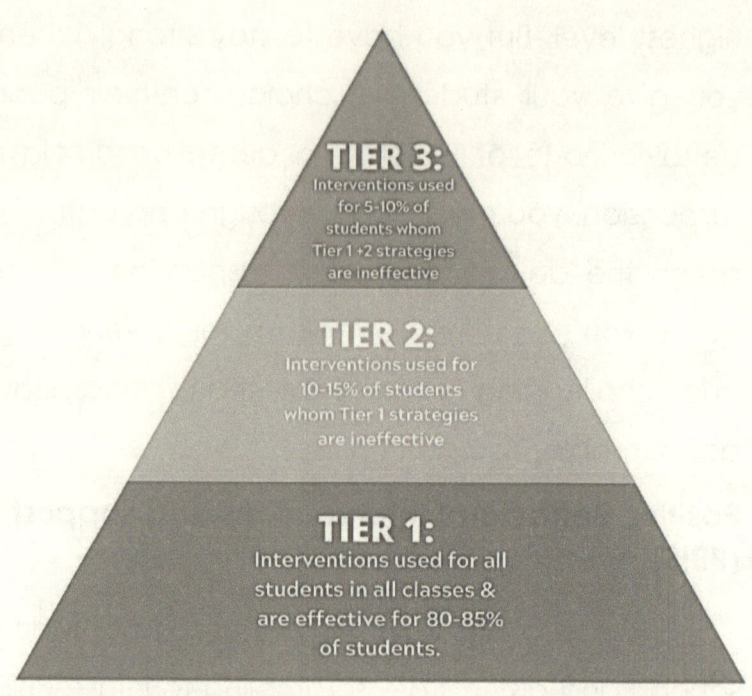

TIER 3:
Interventions used
for 5-10% of
students whom
Tier 1 +2 strategies
are ineffective

TIER 2:
Interventions used for
10-15% of students
whom Tier 1 strategies
are ineffective

TIER 1:
Interventions used for all
students in all classes &
are effective for 80-85%
of students.

Generally, Tier 1 interventions are used for all students. Every single student you have in your classroom should have access to PBIS tools and strategies. There are several explicit strategies for elementary and secondary teachers that can be found on pbisworld.com, but generally speaking, it's a set of positive interventions instead of punitive interventions. For as many years as public education has been a right and not a privilege, there was an attitude amongst many teachers of, "shape up or ship out." If kids couldn't abide by classroom rules, they shouldn't be allowed in the classroom. So, they were assigned a detention and/or a class suspension, or in severe cases an off-campus suspension, and the teacher could resume teaching without the off-task student's nonsense.

This method worked wonders for many years, until data began to show that certain students were being suspended more than others, primarily minority students, students with disabilities, foster youth, English Learners, and socioeconomically-disadvantaged kids. This turned into a giant lawsuit because students in these categories weren't in class as much due to their suspensions and therefore lacked equal access to learning the material in class. This meant they weren't achieving as much as students who were suspended less. So schools and districts began scrambling to find alternatives to punitive disciplinary measures.

Enter PBIS. While there are specific strategies outlined at pbisworld.com, the overall PBIS model consists of looking at each individual student as a whole person. As teachers, we've become bogged down with the pressure of increasing test scores, so we've drastically reduced the amount of positive recognition we give our students for their efforts. You may be saying to yourself, "But I teach middle school" or, "But I teach high school" and, "Aren't they too old for that?" Absolutely not! Think of your own job. I know how hard you work and can definitely say that teaching is one of the hardest jobs on the planet, second to being a parent. We rarely get recognized for our efforts and feel a tremendous sense of underappreciation and hopelessness because of it. I'm sure you can remember specific instances where someone told you how awesome you were doing with your lesson and/or classroom activity. If it affects us as adults, imagine what having that positive recognition does for students.

Notice your students for being on task. Compliment them when they're doing things you want them to do. If you can flood your students with positivity, you'll cut your behavior issues in half. The table below gives you some ideas of how to utilize PBIS to flood positivity in your classroom.

BEHAVIOR	**POSITIVE LANGUAGE**
Student taking notes.	*"Johnny, I love how you're taking notes. That's awesome!"*
Student following rules.	*"Kacie, thank you for coming in and getting right to work on the warm-up."*
Student attempting to do his or her work.	*"Jane, I want you to know that I see how much effort you're putting in. That's so awesome."*
Student saying nice or positive things to other students.	*"Joe, you're so awesome. Thank you for being such a good influence in class."*
Student sitting quietly.	*"Kevin, thank you for sitting quietly."*

Restorative Practices (RP)

Restorative practices are beautiful when implemented with fidelity. So, what exactly are restorative practices? The International Institute of

Restorative Practices defines it as, "an emerging social science that studies how to strengthen relationships between individuals as well as social connections within communities." There are many layers to restorative practices, but we'll cover the most important for managing behavior in this section.

Restorative Practices are all about building relationships. We are a social species, and as such, we thrive when interacting with other people. There are several ways to help build relationships in your classroom, beginning with affective language. Affective language is using a set of words to communicate to your students how their behavior is affecting you, which ultimately helps develop empathy and respect in your classroom. The table below outlines how to use affective language in different scenarios.

SCENARIO	AFFECTIVE LANGUAGE
Student talking during lesson.	*"It really frustrates me when you talk during the lesson. It makes me feel really disrespected."* **OR** *"When you talk during the lesson, it makes it really hard for other students to listen and understand the math. It's really frustrating."*
Student refusing to do work.	*"It makes me really sad when you don't do your work. You are so much smarter than you give yourself credit for."*
Student bullying or being unkind to another student.	*"It really makes me sad when you say such unkind things. If someone said things like that to/about you, how would you feel?"*

In my experience, when I've used affective language, it has helped diffuse tension and helps kids see where you are coming from. One year during lunch, a massive fight almost broke out between several students in my fifth period class. As kids were walking up to my classroom, I could feel the tension. I stopped one student who was particularly heated and asked him what was going on. He explained that everyone was upset with one student in class and that several students wanted to fight that particular student. My initial instinct was to call the office to get administration to handle it, but we had been working on restorative practices all year, so I listened when he said, "Miss Miller please don't involve the office. We can handle this here." My response was, "You know I'm all about handling things in house, but when I have five of you in here right now who can't control yourselves, it makes me feel like I can't handle this on my own." The students who were extra upset began patrolling each other to make sure they were diffusing tension appropriately.

Using RP during conflicts. Conflicts happen. We all come from different backgrounds and with that, we inevitably have conflicts with one another. Using restorative practice techniques, including restorative circles combined with affective language, we can help diffuse conflicts among students. How does this work? When conflicts arise in class, ask questions to get to the bottom of what happened. It is incredibly important to note that *this set of affective questions can be used for minor conflicts in your classroom;* **major conflicts should always be referred to the counselors or administrators.** If you have a minor conflict and would like to diffuse the tension, here is how you do it:

- Talk with each person involved, separately at first, then together, and ask:
  o *What happened?*
  o *What were you thinking about at the time?*
  o *What have you thought about since?*
  o *Who has been affected by this situation?*
  o *What needs to be done to make things right?*

After talking with each student separately, ask each if they would be willing to sit down with the other student to talk it out. Depending on their responses, if you feel you can facilitate a conversation between them without a lot of heated conflict and/or angst, do so, and go through the above questions again. It is vital to prepare each student for the conversation, telling them that it won't be fun or comfortable, but it will dramatically improve the relationship between the students moving forward if they take it seriously.

After each person has had a chance to talk, everyone should come to a set of agreements about how they will interact with each other moving forward. As a vice principal, I would type up the agreements and have each student sign it, but if these are smaller infractions, it may not be necessary. A huge amount of training needs to take place for restorative conferences to be effective, but you can begin using these questions for conflicts that come up in your class to help diffuse tension. I would recommend getting the rest of your class working on something related to the class content, and then take the two students outside for an impromptu restorative conversation. Be sure to keep your door propped open a little bit so you can keep an eye on your classroom while facilitating the conversation.

Trauma-Informed Practices (TIP)

There is new, groundbreaking research on trauma-informed practices, led by Dr. Bruce Perry of the Child Trauma Academy, as well as Dr. Nadine Burke Harris of Center for Youth Wellness in San Francisco, California. Dr. Burke Harris was receiving a huge number of referrals for patients in her clinic in San Francisco, all for Attention Deficit Hyperactivity Disorder (ADHD). When she began looking a little deeper, she saw a pattern with all these kids — that they had been exposed to trauma in their childhoods. With this, their brain development was stunted, and they stopped growing as quickly as they should have, ultimately falling behind their peers.

Dr. Perry developed a list of several strategies to help kids who come from traumatic backgrounds to help bring them up developmentally to match their chronological age. For middle and high school students, part of this includes teaching them the neuroscience of how their brain works.

The following figure shows the brain, and how we can emulate the brain using our hand (more specifically a closed fist). In your hand, your wrist represents your brain stem, your top four fingers represent your cortex, and your thumb represents your amygdala.

**TRIGGERED BRAIN
(Flipped Lid)**

NON-TRIGGERED BRAIN

Cortex
(Four
Fingers)

Brain
Stem
(Wrist)

Amygdala
(Thumb)

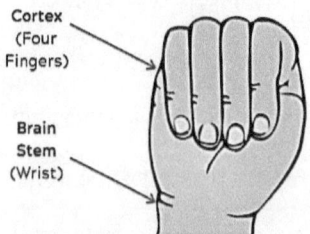
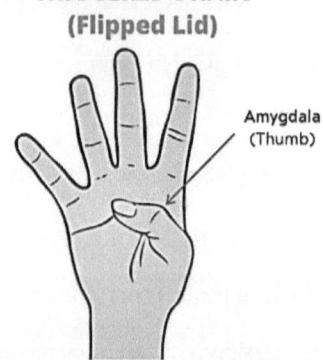

- Student is relaxed and/or calm
- Student can think rationally
- Student can focus on academic subjects
- Cortex in the driver's seat (responsible for higher-level thinking)

- Student is in fight-flight-freeze mode
- Student can NOT think rationally
- Student can NOT focus on academic subjects
- Student is highly reactive and emotional
- Amygdala in the driver's seat (responsible for sounding the alarm)

In this mini-lesson for teens, you explain to them which parts of their hand represent the corresponding parts of their brain, and how when they feel threatened or scared or emotionally out of control, what's happening is they are flipping their lids. What this means is that the amygdala (aka the thumb) is the one driving the bus, which means they're no longer able to think rationally about what's happening. They've entered "fight or flight" mode, and when in this mode, they make really bad decisions because their brain is telling them they're in a life or death situation. This worked well for our ancestors when we were still living in caves and genuinely experienced matters of life or death all the time (aka bears, mountain lions, wolves, etc.), but for the twenty-first century student, these "life or death" situations could be as simple as hearing a sound that reminds them of a traumatic incident in their past.

As a teacher, you likely have kids in your classroom who are triggered all the time. Even kids who have not experienced trauma will be triggered to remember unpleasant events from their past and can also be catapulted into "fight or flight" unexpectedly, causing a raucous in your classroom. So how do you get kids out of "fight or flight" and into their right minds? The chart below outlines different activities you can do with all of your students, traumatized or not, that will help them get out of their amygdala and back into their cortex where all logic, reason, and cognitive thinking lives. When they're living in their amygdala, we won't be able to teach them to solve equations for x anyway. Biology says so.

SITUATION	HOW IT LOOKS (Amygdala Response)	HOW TO DEAL (How to get them into their cortex)	DETAILS
Johnny is triggered by a sound or smell in your classroom.	Johnny disrupts and distracts the lesson, attempting to throw off the entire class.	Take a brain break or have Johnny take a brain break (go outside for a brief walk). Practice box breathing*.	Have "brain break" cards so he can take a quick walk outside*.
Jane has a hard time focusing during the lesson.	Jane is tapping the desk, tapping her foot on the ground, and trying to engage other students in conversation.	Give Jane a stress ball to play with quietly during the lesson. If she insists on tapping, let her tap with a Q-tip so no one else hears it.	Have a basket of stress balls or fidget spinners on hand in your classroom for these types of situations.

Johnny and Jimmy are engaged in a verbal argument, which is escalating to "posturing up" behavior.	Johnny and Jimmy are yelling at each other. They eventually get up and in each other's' faces.	Have one student go outside, call the office, practice "hand breathing" with student inside*.	Deep breathing is a way to help regulate students in "fight or flight" so they can make better decisions.

*Detailed explanations of box breathing, hand breathing and brain-break cards can be found at www.WithHeartProject.com

What's amazing about trauma-informed practices is that they work for everyone, traumatized or not. Think about it this way: teaching and managing a twenty-first century classroom can often feel like trying to herd a bunch of drunk cattle. You have to remember that kids show up in your classroom every day under the influence of something. Sometimes they're under the influence of drugs and/or alcohol. But the vast majority of students are operating under the influence of fear, frustration, impatience, lack of self-worth, anxiety, depression, sometimes trauma, and poverty. It's up to you to figure out what works and what doesn't so you can give all your students equal access to learning and growing in school and in life.

"Why You Always Look at Us Like We're Hella Dumb?"

 Managing your own classroom is like one giant, difficult experiment. You try things. Some work; some don't. You try more things, some of which work, some of which don't, but ultimately one of the most important lessons I learned came from a student who asked me the question, "Miss Miller, why you always look at us like we're hella dumb?" From that point on, I began paying attention to how I was coming across to my students.

I said, "What do you mean?"

He said, "Whenever we ask questions, you always look at us like we're hella dumb."

I looked at the rest of the class and asked, "Do I do that? Do I look at you guys like you're dumb?"

The class unanimously agreed with the other student. I was floored. All I ever had for my students was love and hope and compassion, so this was shocking to me. I could only do one thing. I asked, "Okay, you guys have to show me what I look like when I look at you like you're dumb, because I definitely don't believe you are dumb."

They all started scrunching their faces, so I said, "Okay, I think I know that face, is it this?" I proceeded to make the "hella dumb" face (see picture), and they all shouted, "Yeah, that's it!"

I said, "Oh, my gosh! Okay, I have to explain because I definitely don't think any of you are dumb. I usually make that face when you're asking a question, right?"

They all said, "Yes."

I said, "Okay, the reason I make that face is because I'm concentrating very intently on what you are saying to make sure I understand the question, so I can answer it correctly. It's 100 percent not because I think you're dumb. I think you are all extremely smart."

From that point on, I became incredibly aware of my facial expressions, body language, tone, and anything else they could possibly absorb and misinterpret. Here are the lessons I've learned from my students on:

- **Facial Expressions**: We've already seen in the above example how facial expressions can be interpreted in ways you may not even be aware of. My advice to you is to be mindful moving forward. You may be making a face right now that you don't know you're making, and kids will pick up on it and internalize it like they did something wrong.

- **Body Language**: If you're mad, kids will feel it and see it in your body language, and they will act accordingly. We have mirror neurons in our brain so that when we project a certain quality or characteristic, our students will emulate it. If you show up to class mad or angry or frustrated, kids will mirror you and you'll likely have a very difficult classroom environment to manage

- **Tone**: As with body language, how you speak to your students will make a huge impact on how you're treated as a teacher. Students want to be respected. If you're talking down to your students, using sarcasm and calling students out in front of each other, yelling at students, putting your students down, etc., you may end up with a compliant class, but you'll also end up with a lot of students who have little to no respect for you. I had a student once who refused to do work in her history class because her teacher would constantly put her down in front of the class. He would say things like, "Anne, don't bother doing that assignment. You're not going to pass anyway." Be kind to your students. You can be kind and firm at the same time.

Overall, be mindful, be respectful, be positive, hold students accountable, model empathy, give choices, and you'll virtually eliminate bad behavior.

Social-Emotional Learning (SEL)

Combining all the above techniques and utilizing them in your classroom gives you explicit ways to be a more effective, emotionally intelligent teacher. Studies being released show that "soft skills" (also known as social-emotional skills) are far more important than "hard skills" (also known as academic skills). Many employers nowadays are looking for explicit social-emotional competencies when hiring new employees. With this, it is becoming increasingly important to model and incorporate these skills in our classrooms. Incorporating *Love and Logic*, PBIS, RP, TIP, and general communication skills into our classrooms will give our students the "soft skills" they need to be professionally competitive moving forward.

Oftentimes people think SEL is teaching all about emotions and touchy-feely stuff, but really, it's teaching kids to be well-rounded, respectful human beings that can interact with each other in positive and productive ways.

 Have a specific behavior-related question, comment or concern? I want to know what you think!

Head to **withheartproject.com/contact** so we can chat!

Chapter 5: Maintain Positive Admin Relationships

When I first started teaching, I questioned what vice principals did all day. I thought they literally sat in their offices twiddling their thumbs while I worked my a$# off with thirty, sometimes forty, kids every hour. The interactions I had with vice principals were almost always pleasant, but as time went on and I needed additional support and didn't get it, I was pissed. I thought to myself, *What the hell? I'm busting my butt to give these kids a great education, but Johnny keeps interrupting my lesson and won't let me finish. The vice principals tell me to contact them and let them know when I need help. But when I do, no one comes.* It was infuriating!

Case in point: When I caught a kid in my class with weed, the vice principal told me, "It's your word against his."

We talked in the last chapter about the importance of holding kids accountable, so when this happened — and the kid got zero consequences — I was *fuming mad*. In my head, it went something like this: *If you don't give the kid consequences, then all other kids on campus will get the message that they can bring weed to school and nothing will happen to them. How can they be okay with sending the absolute wrong message?* Same concept goes for disrupting class, student fights, bullying, and any other ways kids can break school rules. If you don't hold them accountable, it's just going to get worse.

Here's another example: I was a high school math teacher for the vast majority of my career, although I did have brief stints in AVID and CTE. I had my one classroom, four walls — that was it. The most danger kids were in was with each other and potential fights breaking out. There really wasn't anything in my classroom from which kids could get physically hurt. That all changed when I was hired to create a CTE pathway, write the courses for the pathway, and stock my new "classroom" with a boatload of power tools to support the curriculum. The classroom space was amazing, but I was no longer responsible for just four walls — I had like sixteen. This new CTE area had four separate learning spaces: the classroom (like I was used to with a whiteboard, projector, and bunch of student desks), a computer lab (consisting of chairs, desks, and computers), a hands-on lab (consisting of lab tables, stools, lots of tools, and materials for hands-on building), and an outdoor area (consisting of larger power tools where kids could build larger projects). It was amazing.

So, one of our first assignments in this space was to design recycling bins to put around campus. We started by using architectural rulers to make scaled drawings of a bin that would be engaging and essentially *make* students *want* to recycle. I know what you're thinking. *Doesn't everyone just innately want to recycle?* Not high school students at a Title 1 school. That's the *last* thing on their minds. Anyway, before kids could proceed with the building aspect of the project, they had to pass their safety tests, get their scaled drawing approved by me, and show me a breakdown of what materials they planned to use and the dimensions of each piece that needed to be cut.

Team B was the second or third team to get going on constructing their bin. They were using plywood to construct their bin, so part of their team went outside and set up their materials, got the saw, and began cutting. I continued monitoring the construction; then a student inside the lab space asked for help. I went inside to help the student when another student from Team B rushed inside and said frantically, "Miss Miller, Jackson got hurt." I stopped what I was doing and intercepted Jackson as rushed inside, looking like he was in immense pain, "Okay show me what happened," I said, "show me where you got hurt."

He showed me his hand, and he accidentally cut one of his fingers with the circular saw outside. He didn't cut his finger *off*, but when he showed me his hand, I could see he cut deep enough to expose bone. I was calmer and more collected than I expected to be in such a situation, so I jumped into "emergency management mode." I got a bandage and wrapped up his fingers, then immediately called the office and asked for help, which thankfully came right away. Jackson was rushed off to the hospital, and I went back to the thirty-one other students in the lab who were all just staring at me, waiting to see what I was going to do. Class was almost over, so I only had a few minutes to fill before I could take a breath and go into panic attack mode. I stayed calm, cool and collected, and explained what happened and that Jackson was now getting help — but I also talked about the importance of following proper safety procedures. Over the next day or two, I would learn about how proper safety protocols were not followed after I left the outdoor cutting area to help another student.

This event happened in September or October. Jackson ended up being fine. He came back to school and continued with his education. He was a football player and was no longer able to play football with this injury, but overall, we finished out the school year with no more injuries. Everything seemed to be looking positive for the future. Until I was called into the principal's office two weeks before the end of the school year.

When I went into the office, both the principal (who happened to be a first-year principal) and the CTE director from the district office were waiting for me. I immediately got a bad vibe, but I sat down. I looked across the table at my principal, and she looked defeated. She was visibly nervous, and said, "I don't know where to begin." The CTE director interjected and said that I would no longer be teaching the CTE class — the CTE pathway that I had built. I honestly don't remember the exact reaction I had, but it must have been visible because they both asked if I was okay. I think I replied, "No, but I'll try to be." I left the office totally destroyed. I headed back to my four classrooms and was greeted by the CTE department chair, who could tell I was not doing well. He asked what was going on and I said "I'm out. I'm not teaching CTE next year." His face dropped and he angrily asked, "Why?" I said, "Because of the injury that happened." He shook his head and was **not** happy. I went into my classroom and cried.

I worked for two years to build this pathway. Before I began, this pathway was just a thought. Then it became a dream that I then made a reality. I spent *so much time* designing the classes, which ended up being around 120 single-spaced pages of course descriptions. I researched equipment and tools to support the curriculum. I completed purchase orders for $150,000 worth of equipment, tools and supplies, and inventoried the curriculum as it came in. I set up the entire lab space, which was formerly being used for storage. Then I took everything and made it into a viable daily curriculum with activities to be used in the class. I saw kids transform. Kids who were a pain in the a$# in my math classes transformed into the leaders and project managers in the CTE class. Kids were inspired. They were doing work to give back to communities in need. All of it went to sh@$ the day I was told I would no longer be running it. I was destroyed.

I'm telling you this story not because I love rehashing it, but because there is value here. The day I got this news, I'm pretty sure I went home and drank an entire bottle of wine and cried. Even now as I sit here typing these words, I'm getting flushed and frustrated, even though this was more than three years ago. I couldn't understand why that decision was made. I was pissed at the CTE director and pissed at my principal. I thought she should have had my back. I thought she should have supported me and fought for me to keep that pathway. But she didn't. She was a brand new principal and felt defeated. I always said to myself, "Ya know what? When I get into administration, I'm *always* going to support my teachers, no matter what. They have the most important job in the world, and my ability to do my job doesn't need to make their job harder."

So, the following year when I was back to full-time math instruction, I worked to get my Master of Science in Educational Leadership and my administration credential. I busted my tail to secure a vice principal position for the following year; I was on my way, ready to defend teachers to the death — until I realized that administrators only have so much power over large decisions made day in and day out.

I absolutely loved being a vice principal. Well, mostly. I did defend teachers to the death, but on a few occasions, I came to see how much politics interfered with my ability to do so. The year I was a vice principal, I worked at a middle school located right next door to its feeder high school. During the beginning of the school year, this high school experienced an onslaught of racism that one of its students ignited by making a video of herself saying, "I hate black people," and posting it on social media. It ended up on the news, meaning how school and district leaders handled the situation would be highly scrutinized. While it wasn't our school, the fact that a library and bus loop were the only things separating our campuses meant the district wrapped us up in the same category.

Our middle school administrative team took a deep sigh of relief that we weren't going through this racist madness — and then we were going through it. Two incidents stand out in my mind during this time: In one, we, as the site administrative team, took matters into our own hands; in the other, we had to wait for instruction from the district office.

In the first incident, two female students painted their faces black and wrote the word "ni@#$#" on their foreheads, then posted a photo to social media. Because we were all wrapped up in restorative practices at that time, we, as an administrative team, decided to issue suspensions for these kids but reduce the number of suspension days if they participated in a formal restorative conference. We handled it — but apparently not in a way the district office wanted. From that point forward, we were told to alert the district office of any race-related situations that arose and wait for their instructions on how to handle them.

In the second incident, a male student stood up on one of the lunch tables in the cafeteria, made the Hitler hand motion and said,

"Hail Hitler" in front of other kids. The campus security guard reported it to one of the other vice principals, and our principal called the district office, which told her to conduct the investigation but hand down **no** consequences until the Superintendent and his Cabinet determined *what* the consequence should be. It took two or three days for the district office to get back to us on this; meanwhile, the offending student is at school and attending classes, and it appears to everyone on campus (students, teachers, *everyone*) that he hasn't been given a consequence for this behavior. All the teachers were pissed. In the same way I felt I got no support from administration when I caught the kid with weed, teachers felt like administration wasn't doing their job and we were just letting kids throw up Hitler signs without consequence. *It was a problem.*

I tell you this because I think it's important to shed light on administrator's jobs. Until you've been an administrator, you have no idea what it's like. A typical day for me when I was a vice principal could include any or all of the following:

- Investigating incidents **(one and a half hours+ *per incident*)**
- Interview all students involved in each incident **(15-30 minutes per student)**
- Get written statement from all students involved in each incident **(5-10 minutes)**
- Determine consequence **(5-10 minutes for clear cut situations, more than that for cloudy situations)**
- Enter incident into Student Information System **(5 minutes)**
- Fill out appropriate paperwork for consequences **(5-10 minutes)**
- Call all parents **(15-30 minutes per parent)**
- Observing teachers **(20+ minutes per teacher for informal observations)**
- Formal observations **(one hour per teacher)**

- Writing evaluations **(one-three hours per teacher per year, four-five hours per evaluation)**
- Meeting with teachers for evaluation cycle **(15-30 minutes per meeting)**
- Supervising during lunch **(30 minutes per lunch, two lunches per day)**
- Supervising during passing periods, before school and after school **(one and a half hours+)**
- Meeting with students who are struggling with mental health concerns and potentially contemplating suicide **(30-60 minutes)**
- Call School Resource Officer (SRO) to do evaluation of student **(5-10 minutes)**
- Intervene during health crises that come up **(15-30+ minutes)**
- Following students to the hospital when necessary & waiting for parent

to arrive at the hospital **(30-60+ minutes)**

- Facilitating restorative conferences **(45-90 minutes per conference)**
- Interviewing all participants ahead of time for conference **(15-30+ minutes per participant)**
- Putting out fires that come up in various classrooms or places on campus **(15 minutes + per fire)**
- IEPs, 504s, SSTs **(30-60+ minutes per meeting)**
- Attendance-related meetings **(30-60+ minutes per meeting)**
- Expulsion hearings **(45-60+ minutes per meeting)**

It's a lot, right? And it doesn't even include district-level trainings and adjunct duty supervision. The point I'm trying to make is that administrators are *busy*. I'm not saying they're too busy to support you. In my professional opinion, that should be their first priority no matter what.

You are on the front lines with students, so short of physical safety issues with students, you need to be their priority all the time. Point blank, no exceptions. That being said, they are busy, and they're held to a standard of communication that is far beyond (and more ridiculous than) that of most teachers. What do I mean by this?

Administrators are the face of the school. When the sh@# goes down, you as a teacher may feel as though you're in the spotlight. But truthfully, an administrator's job is to intervene when things get ugly, therefore placing them square in the spotlight. They must be *incredibly* careful what they say, how they say it, what they do, and how they do it. Because if they aren't, many parents and community members are out there waiting to pounce and slap on a lawsuit. Of course, as teachers, we need to remain professional. But when it all is said and done, people really aren't looking at you. They're looking at vice principals, principals, and district office administrators.

When I look at it from the perspective of having lost my CTE job, I was pissed that I wasn't supported (or that it didn't *feel* like I was supported). I genuinely believed that my principal should have fought to keep me in that position. It wasn't until much later that I learned Jackson's family filed a lawsuit against the district, claiming I was a negligent teacher. While I don't know all the details, I would imagine that from Jackson's parents' perspective, had I been doing my job, I would have intervened properly, and Jackson wouldn't have gotten hurt. So, in the end, my job was a bargaining chip for Jackson's family. Perhaps they asked for less money if I was removed from my position, or maybe they said they'd drop the lawsuit altogether. When we start getting into situations like this, it really **is** out of site-level administration's hands. Those of us who really need to find someone to blame can run it up to someone at the district office. But then truthfully, they're at the mercy of lawyers and lawsuits. And we already know that funding in education isn't exactly flowing.

I know it feels easy to point fingers at administration to make ourselves feel better about what's happening in our classrooms. "If administration would just do their job, I could do mine." I hear that a lot. But what I've come to understand — and what I'm trying to get you to also understand — is that administration *is* doing their job. They're just doing it in a *really* screwed up system.

Here's another story to further illustrate this concept.

My last year in the classroom, I was at a middle school site that experienced a leadership change in which the principal who held that position for ten years took a job at the district office. The incoming principal had a very different approach to discipline, and as such, wanted to implement more restorative practices, less punitive discipline.

This school year was *nuts*. In my thirteen years in education, I had never experienced anything like it. Complete bananas. It wasn't uncommon for two to three fights per day to break out; one day it got up to eight fights. *In one day*. The first semester was complete turmoil for all of us teachers. We didn't feel supported by administration, some of whom would straight up say to teachers, "I can't support that," on various issues. We were feeling beat down and like the kids were in control. They didn't have consequences and felt they could get away with whatever they wanted.

So, right around January following that very trying first semester, I was told that my colleague (a long-time, veteran teacher at this school) was walking in the hallway and kids were in the way; he just needed to get by so he could be on his way. So, he said, "Excuse me." Kids didn't move. The teacher got frustrated and said in an angrier tone, *"Excuse me."* At this point, I don't know if he nudged a student or what exactly happened, but the student decided to turn around and say to the teacher, "That was assault." That afternoon, the teacher was removed from his classroom and put on paid administrative leave while the incident was being investigated. He never returned that year and teachers were *pissed*. They felt like the teacher wasn't being treated fairly, that he wasn't getting supported by admin, etc. The truth is, I still don't know the outcome of this incident, though I imagine it's still being investigated.

In the moment, however, I, having had prior administration experience, tried to calm the waters. I explained to teachers a protocol exists for any time an allegation is made against a staff member, and that administration was following that protocol. Right, wrong or indifferent, they were following the protocol as set forth by the district office. So, from the outside, his removal looks like the administration isn't supporting him when, administration *has to* follow this protocol because unfortunately there *are* teachers who break laws and do highly inappropriate things with students. If you're unaware of any of these situations, look on the news. There are plenty of these stories out there.

I told my fellow teachers that at this point, it was out of our principal's and vice principal's hands and that district office administrators would make a decision. So, let's look at this situation's possible outcomes:

OUTCOME 1: ADMINISTRATION ALLOWS THE TEACHER TO COME BACK.

In this potential outcome, everything carries on like normal (or as normal as it can be given what just happened). But what does this look like to the teachers and student population and their parents or families? It ultimately sends two messages:

- **Message 1**: Teachers can do whatever they want with no consequences.
- **Message 2**: The principal and vice principals and administrators take the word of a teacher over the word of the kid.

OUTCOME 2: ADMINISTRATION DOESN'T ALLOW THE TEACHER TO COME BACK.

In this outcome, a replacement teacher is found, and everything carries on as normal as it can given what just happened. What does this look like to the teachers and student population and their parents/families? It ultimately sends two messages:

- **Message 1**: Students can do whatever they want with no consequences.
- **Message 2**: The principal and vice principals and administrators take the word of a student over the word of the kid.

It's a lose-lose situation no matter the outcome. Before I started as a vice principal, the Director of Secondary Education said to all of us vice principals, "Just imagine that a news crew is following you around on campus every day. Would they have anything to report? That's what you need to be thinking about in terms of how you conduct yourself."

Administration Empathy

We're in a different time, a different era. Kids with cameras and smart phones are everywhere, and many are looking to film the next big event that will put them on the map. Unfortunately, some take this idea to extremes. Take the demeanor on our campus after the "excuse me" nudge turned assault charge — the kids created a game called "Teacher Gone" where they teamed up, deliberately provoked teachers and filmed it in an attempt to get the teacher fired.

Bottom line is this: As teachers, we can point the fingers at administration all we want. But this is taking precious time away from things you could be doing and creating to better teach your students how to be respectful human beings who wouldn't want to get a teacher fired in the first place. This isn't an "us versus them" issue. It's not a "teachers vs. administration" issue. *It's a **societal issue.*** So what can you do to make your campus a more respectful place? Here are a few tips to get you started:

- Administrators are people just like you. They have their own stuff going on at home and not everything is about you.

- Sometimes administrators aren't the ones making the final decision. If you're unsure, go ask them about it! If you're unsure about what's going on with a particular student, teacher, or issue, go ask *in person.* It doesn't need to be a formal meeting; if you see your administrator in passing, informally ask any questions you want; if they can give you information, they will.

- Sometimes administrators can't give you any information. It's not because they don't *want* to, it's because legally they *can't*. In an open investigation, if they start telling every teacher on campus about details of the investigation, not only does this potentially taint their investigation, but they're violating privacy and confidentiality laws.

- Some administrators think the rules they must abide by are as stupid as you do. Take formal evaluations for example. When I was trained to do formal evaluations, I was told I was *not allowed* to rate any probationary teacher higher than the "you're doing good but need improvement" category. Sometimes I had rock star teachers and wanted to just tell them that, but the bureaucracy was tying my hands.

- Try not to take it personally. I'm not going to lie, there are some corrupt administrators out there, but most of them are just trying to do a good job. Give them the benefit of the doubt.

- Start trying to practice administration empathy. They may not have thirty kids to deal with every hour, but they do have the hardest thirty kids in the entire school to deal with every day, every hour...sometimes one at a time, sometimes several at a time. Take comfort in the face that you have *a lot* of amazing students and do have the ability to call for an airbag if Johnny acts up for the sixty-sixth time in the last three weeks. Who is your airbag? Your administrative team. They don't have an airbag. Keep that in mind.

Teacher Empathy

I don't think it would be right to end this chapter without giving some tips to administrators when working with teachers, so this is my gift to you:

- Teachers are people, just like you. They have their own stuff going on at home and sometimes just need a little extra help. Honor that.

- Please, please, **please** try to remember what it's like to be a teacher. Yes, we realize you are our airbag for the most challenging students. We deeply thank you for that. But please try to remember that not only do I have challenging Johnny in my class, I've also got twenty-nine others that all present their own challenges. I have to manage *all of them* simultaneously while also teaching them to be proficient at solving equations for x.

- If I'm asking for your help, I genuinely need it and some guidance as to how to better work with my students. Instead of just not showing up or not dealing with it, help me! If you can show me how to interact with Johnny in a more effective way next time, maybe I won't have to involve you!

- Try not to take it personally. Sometimes as an administrator, you want to pull your hair out because "Mr. Jones can't handle his class." Take my advice: If your teachers are struggling to manage their classrooms, perhaps that's a clue that *they need more training* in how to handle their classrooms and the challenging behaviors in them. Yes, I know, budgets are tight, time is of the essence — but find the money, and find time. You'll all be better off for it in the long run.

- Please try to be mindful and have teacher empathy. You were a teacher once, and you know its challenges. Honor that, honor your teachers for all the effort they are putting in daily. Maybe even throw in a little positive recognition from time to time. Research shows that *everyone*, young or old, loves to be recognized. If you must mark me at a low level on my teacher evaluation because those were the instructions you were given, please just say that. You can say, "Hey, I think you're doing a great job, but the district office tells me I *have to* rate you here. But you're doing awesome."

- If I ask you for information about a student and/or incident, please don't blow me off. I'd feel much better if you just said, "I really want to loop you in on what's happening, but I can't just yet. Can you check back with me in a couple days?" That response is better than no response.

- Finally, if I call the office and need your immediate support, please show up. There is nothing more destructive for our relationship and the culture on campus than feeling like we're working against each other.

- We're all on the same team. We all have a common goal. We all got into this profession because we love kids. We love helping kids realize their full potential, we want to help students succeed. If we stop pointing the finger at each other and start practicing empathy for one another, we can move mountains and transform kids' lives together. Let's do this.

Have a specific administration-related question, comment or concern? I want to know what you think!

Head to **withheartproject.com/contact** so we can chat!

Chapter 6: Push Through the Politics

If the bad behavior and seeming lack of support from administration doesn't make you want to run as far from the profession as humanly possible, the politics just might. And for me, the hardest lessons to learn in the education field were by far the political ones, especially since I could do nothing about the several political situations I experienced, like:

- A student being transferred into my already-cramped geometry class (bringing the class size to forty-three students while other geometry classes were as low as fifteen students), because the student's mom blamed all previous eight teachers for her son's failing grade.
- A teacher next door to me sleeping during instructional time, while I was busting my tail to provide kids with a great education (don't get me

started on the whole teacher tenure thing).

- Grievance issues by colleagues, followed by petty communication and behavior.
- Football coaches asking for academic favors for their players.
- The whole "college for all" mentality (maybe some kids genuinely don't want to go to college (or aren't developmentally ready)).
- Being given a goal to double proficiency on the Algebra 2 STAR Test (now known as CAASPP, Integrated Math 3).
- Being judged as a teacher by your students' test scores.
- Being expected to bring all kids to proficient levels in Algebra 2, despite having students who literally can't add 2 + 3.

- Being removed from certain teaching assignments because of lawsuits and/or parental pressure.
- Getting ultimatums and threats from parents if their kids didn't pass.
- Kids being caught with drugs and/or alcohol and having virtually no consequences.

This list sounds daunting — and believe it or not, there's a whole host of other issues — but it's the unfortunate reality in today's education system. Teachers face these very issues every day. Administrators face these issues every day, too. In fact, my most painful lesson in educational politics happened during my time as an administrator.

I wanted to be a vice principal from the time I was in my first or second year of teaching. When I created my CTE pathway, my dreams of being a vice principal went out the window. I thought I had found my calling, and my calling was being a CTE teacher. Then Jackson got injured and I was removed from my position. The path that I painstakingly built in my CTE pathway was no longer in front of me, so I went to the next best thing: administration. As I mentioned, I wanted to be a vice principal ever since my first few years of teaching, because I consistently saw administrators "handle" situations that didn't end up getting handled. In my head, I knew 100 percent I could do a better job than many vice principals. So, when I was given an opportunity to test my skills in administration, I was *thrilled*.

After three rounds of grueling interviews, I was chosen out to serve as a vice principal at a high-performing, top-of-the-line middle school in a local school district. At this point, my only experience in administration was as a vice principal substitute shortly before I was hired. Meeting the principal and rest of the administrative team made me feel like I was home. The newly hired principal, Katie, couldn't sit in on the initial panel of interviews, but she was part of the final interview where I was given scenarios to test my ability to handle high-level conflict, high-pressure situations with angry parents. I passed the test and was offered the job.

The summer before I started, I ran into Katie at a café near the school. When she saw me walk in, she immediately came over and gave me a hug. I felt so incredibly welcomed. After meeting with her, I made my way back to the actual school site and was shown my office, introduced to the office staff, and felt even more welcomed. I felt like I was in my professional home. Everyone kept telling me how excited they were that I was on the team, and it just made me feel like this was my new path. All the devastation I felt from being ripped from my CTE pathway faded away. This was my new path, and this was my new home.

I couldn't wait to get started. The school year began relatively slowly, so the administrative team and I worked on some tasks to "get ahead." I created an informal teacher observation walk-through form so teachers could get immediate feedback on pop-in observations, and I started looking at and meeting with the activities director to discuss weekly advocacy lessons. I did some informal teacher observations, along with all my other supervisory duties (morning supervision, passing period supervision, morning break supervision, lunch supervision, and after school supervision).

Around 1,400 11 to 13-year old students roamed the quad and hallways several times per day, and as we all know, kids need supervision. Luckily others were on campus to help with that: the campus security teams. It was up to us — campus security (considered classified staff) and administration — to make sure all the kids were being safe. Part of my responsibility as an administrator was overseeing various classified staff, and I had the misfortune of overseeing a dysfunctional duo of campus security guards who had to be watched carefully for fear they would kill each other. I was told before I showed up that I was being given these two security guards so a "fresh set of eyes" could work with them. I started slowly by getting to know each of them as individuals and observing to make sure they were monitoring their areas of campus when they were supposed to.

Something you should know about me is that I was put on this earth to serve and empower people. I have spent thirteen years working with students and staff members on school campuses, helping them find their innate strengths, and then giving them the tools needed to make great strides in life. Aside from being a good parent, this is my mission in life. So, in getting to know these security guards — to help not only empower them, but also help them get along with each other — each would undercut the other in private conversations and make their intense hatred for one another known.

Then during these get-to-know-you sessions, things took an awkward turn: One of the security guards, a younger, attractive man named Brian, started saying inappropriate things to me. One that I can't seem to forget is, "Oh, the last thing you thought about before going to bed was me, huh?" My guard went up and I began to tread very lightly.

Those of you in administration know that it's very easy to get holed up in your office for the entire day doing investigations and meeting with students and parents. If an opportunity presented itself, I would sometimes sneak out of my office and take a short five-minute walk around campus. Brian was often sitting on the golf cart around campus. One afternoon in an effort to get to know him better, I struck up a conversation about life. Brian was talking about how he was trying to get the "foundational" pieces of his life in order. I asked, "Like what?" The rest of the conversation went something like this:

> **Brian**: "New apartment, new girlfriend."
>
> **Me** (relieved he mentioned his girlfriend): "Hey, can I be honest with you about something?"
>
> **Brian**: "Sure."
>
> **Me**: "I'm glad you feel comfortable enough with me to bring up your girlfriend. I'm really looking forward to us being friends."

Brian: "No problem. What did you want to be honest with me about?"

Me: "That. What I just said out loud."

Brian: "Oh, I thought it was going to be juicier than that."

Me (awkward): "What do you mean?"

Brian: "Well, before you brought up the friend thing, did you always have a sterile view of me?"

Me: "Well you're an attractive guy, but I'm married, and we work together so... What view do you have of me?"

Brian: "100 percent professional."

Me: "Why did you ask me that question then?"

Brian: "I dunno."

I had gone from having my antennas up to being fully 100 percent uncomfortable. Fortunately, I got a call on the radio that the person I was supposed to meet had just arrived, so I headed back in and had my meeting. Once the meeting was over, I went right next door to the other vice principal's office. Her name was Valerie, and she was also a relatively new administrator. I told her about the interaction I just had with Brian and she looked shocked. I expected her to say something along the lines of, "That is *not* okay. You need to go report this to Katie immediately." But instead, she said, "Why doesn't he say stuff like that to me?" I was now shocked and even more uncomfortable.

I did end up reporting the situation to Katie, and Katie's response was, "Maybe he was just kidding. He does have a weird sense of humor. Do you want me to talk to him?" I didn't really know what to say, but I did know that I had this deep-seated desire to show that I was a capable new vice principal that she could trust to deal with this situation. So I said, "Please don't talk to him, I'd like to handle this myself." And she let me.

That ended up being a terrible decision.

Over the next couple months, nothing got worse. In fact, things were pretty stable. I continued interacting with him and building a professional relationship, and true to form, attempting to help him professionally. I included him in restorative circles and accompanied him on home visits for struggling students so he would have an opportunity to build his resume. Things were fine, or so I thought.

Then things came crashing down in my personal life and I withdrew from everyone at work, including Brian. About a week after I withdrew from everyone, Brian initiated a meeting with Katie and told her that I sexually harassed him and that he wanted to file a formal complaint. He began getting extra creepy from that point on. He stared at me intently from across campus and refused to help with students when I was the vice principal in charge. Katie called me into her office and told me that he was filing a formal complaint against me. I'm pretty sure my response was something to the effect, "Are you kidding me?!" She wrote me a long-winded e-mail stating the nature of the complaint and allegations being brought against me. This was November 2017. She instructed me to have minimal contact with him moving forward, and I continued doing so — but now it was because she was telling me to rather than because I was withdrawing from everyone.

The next couple of months came and went. I proceeded to kick a$# at my job; I handled discipline issues and conducted informal and formal observations and evaluations of my staff. I was constantly being told by teachers that they were so happy I was their vice principal, that I was so supportive, etc. Even Katie would tell me in passing and informally that I was doing a great job. So, nothing — and I mean *nothing* — could have prepared me for my vice principal check-in meeting in January 2018.

It was a Friday, and Katie snuck our meeting in just before my scheduled phone meeting with my master's degree advisor was supposed to start. I should've known something was off since our normal check-ins were typically Wednesday mid-day. We talked a bit about my master's degree and accompanying project, and then her tone and body language changed, and she said, "I have to have an uncomfortable conversation with you."

"Okay," I replied. I was terrified. The rest of the conversation went something like this:

Katie: "I am not recommending that your contract is renewed for next year."

Me (stunned): "Okay, can I ask why?"

Katie: "I can't give you a lot of details right now, but as I work through your evaluation, I'll be able to give you more information. It's not personal, and I want you to know it has nothing to do with what happened with Brian."

Me: "Okay — I am a really hard worker and would really like to know what I'm not doing right so I can improve in the future."

Katie: "That's definitely a strong suit of yours — you're very reflective. Two things that come to mind are that advocacy lessons sometimes come out late and teachers don't know if they're happening. Or that if it's lunch time and you have a student in your office when there are 700 kids outside that need supervision. Those are the two that come to mind, though they're relatively minor (she gets visibly nervous). It sounds like I'm trying to make up reasons. I know this probably looks like a blemish on your record, but ultimately you just weren't ready for this job. It's just not a right fit."

I stared at her blankly, I didn't know what else to say.

Katie: "Well, I know you have your meeting with your advisor, so I won't keep you any longer. I expect that you'll have questions as you process this, feel free to ask anything you'd like. Work with human resources to see if there's another position for you in the district."

I left her office. And I was destroyed. Again. But I composed myself and called my master's degree advisor. We talked about my research for no more than about fifteen minutes, and I gathered up my belongings and headed home.

The drive was eerily quiet. I didn't have the energy or capacity to turn on any music. I just sat there in silence the entire drive home, in a complete daze, a mix of mostly shock, but a little bit of anxiety, outrage and frustration. When I got home, I didn't say hello to my husband, I simply walked in the back of the house and stood there in the bathroom, in a daze. My husband followed me and cheerfully said, "Uhhh, hi!" I don't think I responded, so he again said, "Hi. What's up?" I explained to him that Katie wasn't renewing my contract for next year.

He got angry, "WHY?"

I started to cry and said, "I don't know."

He said, "I certainly hope it doesn't have anything to do with this Brian bullsh@#."

"She said it didn't." I said.

"Then why?" He asked.

I began to describe that wacky meeting I just had, where she couldn't really tell me why she was firing me, but that she just was. And I remained in shock and began going through the stages of grief over the next several months. I never did get answers, but I did begin my mission of documenting everything, and I mean *everything*.

I literally created a Word document and just began typing, starting at the beginning — from Brian's very first inappropriate comment. When all was said and done, I had thirty-eight pages of documentation, as well as a list of several employment laws that were broken, and board policies and education codes that were violated. I finished out the year with as much grace and dignity as I could. I showed up on time, worked my tail off, cared for and loved my students, continued to empower them to be their best, worked with my teachers and classified staff to empower them to be their best. I was doing a pretty good job until May rolled around. At this point, things were becoming increasingly contentious on campus. Brian began refusing to help me with things that were part of his job, he began interacting with students in an aggressive and inappropriate manner, and I continued to do my job through all of it, picking up Brian's slack, until one day after school at the bus loop. There were two students who missed their bus and needed to cross the street so they could catch it on the other side of the street where it stops to pick up high school students. We had a strict rule, for safety reasons, that middle schoolers could not cross the street where the crosswalk was when there were buses. It was too dangerous. They approached me and asked, "Miss Miller, can we please cross the street to catch our bus?" I said to them that they couldn't cross at the crosswalk for safety reasons, but they could walk around the long way by the library. They were on their way and I continued supervising the rest of the students.

Out of the corner of my eye, I saw Brian enraged and going across the street to drag the two students back onto the middle school side of the street, through the dangerous crosswalk no less. I watched from a distance as he got in their faces and they ended up walking off to the office frustrated. Given everything that had occurred with Brian and I earlier in the year, I didn't do anything with that information and went straight home after I was done supervising.

When I arrived at work the next day, one of my fellow administrators asked me for a statement of what happened at buses the prior day. I gave him a confused look. He said, "One of the students' parents called and left me an irate message saying that a 'security guard' wouldn't let him cross the street, and as such he missed his bus and she had to come pick him up." The parent was rightfully angry. I typed my statement and sent it to my colleague. Done.

Right around this time, Katie and Valerie, the other female vice principal, began teaming up against me, calling me out on the radio to make it appear that I wasn't doing my job. They also started including Brian in high-level administration and leadership meetings and having him attend all after-hours functions (adjunct duties for me). So, when I discovered he would be supervising Open House, I asked Katie if I could bow out of the event. She said no. I asked her if Brian was going to be there, and she said she didn't know. I was at my wits end. Earlier in the year, the non-corrupt vice principal colleague of mine urged me to contact the Union president, but I didn't take his advice — until now. I told the union president the situation and that it was becoming increasingly difficult work in such a hostile environment.

Katie must have gotten wind of this because the following day, she called me on the radio and told me to meet her at my office. She had a letter in her hand and was putting me on "paid administrative leave" due to my "unprofessional conduct." I was instructed to clean out my office, turn in my keys and vacate the premises immediately. I would not be allowed to set foot on any district property while the investigation was happening.

What.

The.

F@#$.

I immediately called the union president, at which point he advised me to get a lawyer and gave me a few names. When it was all said and done, I had four to five separate lawyers telling me I should file a lawsuit. So, I started the process by filing a formal complaint with the state of California. I say that like this all happened immediately, but in reality, it took me a good year to realize that while I did have a case and would likely win, I didn't want to put myself through all of that. I had created an organization to work alongside schools and districts instead of within them, and it was going well. I was gaining traction and seeing amazing results with students. I wanted to focus my efforts on that. So, I did.

Okay, that was a long story and you may be wondering why this fits into the "Push Through the Politics" chapter, so let me explain. It turns out that Brian was having an affair with Valerie and was also Katie's "lookout" on campus. When I came into the picture and Brian began hitting on me, Valerie felt threatened. When I started to see all the laws being broken and board policies and education code violations, Katie felt threatened. She had the power to let me go, so she did.

One of the most political and detrimental aspects to the education system is the way people are hired and let go. I became tenured in three separate school districts and was, by all accounts, an exemplary teacher. I earned a "Teacher of the Year" award and was selected to create high-quality curriculum for an innovative education company along with 99 other highly qualified individuals from around the world. I followed the rules. I did my part. I worked my way up, and in the end, it didn't make any difference because I was a probationary employee. Katie never did one single formal evaluation of me. She was supposed to do a minimum of two formal evaluations during my time as probationary vice principal. But she did zero, and I lost my job. And I wasn't the only one who succumbed to her wrath. She also tried to fire the other non-corrupt vice principal, but he was hired under a special contract where he had tenure in the district, and she therefore couldn't throw him out the way she did me. She also fired a hard-working, first-year math teacher at the school who busted her tail to get a whole second credential to help take on classes that a history teacher couldn't take.

The point I'm trying to make is that there are things out of your control as a teacher. I truly hope that you won't get wrapped up in any high-pressure, political situations like those I got wrapped up in, but if you do, you must be able to manage them with dignity and grace. Here are a few ways I managed to survive during this incredibly challenging time that tested me to my core:

- Remember why you show up to work each day. I'm guessing it's not to show off to your administrators what a bada$# teacher you are (and yes, you are a bada$# teacher if you're reading this book). You come to work each day to inspire and empower the students in your classroom. Focus on that.

- Take time to slow down and breathe. If things get to be too much, just stop. Take a break. Maybe even share some of your frustrations (without divulging confidential information) with your students. I know this sounds scary and counterintuitive, but if you show your students that you are going through difficult times, that will build an amazing connection that will just strengthen your ability to teach and reach your students.

- Write down the things that are bothering you! When I began incorporating social-emotional wellness curriculum in my classes, I would sometimes have "snowball fights" with my students. I'd ask each student to write down something that was frustrating them. They were to write with as much passion and anger and angst as they needed to, to make them feel physically better. Then we would either wad the papers up (snowballs) or tear the papers up into tiny little pieces (snowstorm) and throw them at each other. Students would tell me how much this helped them get it out of their head so they could focus better. Maybe this will help you, too!

- Move your body! Whether it's taking a brisk walk, jogging, stretching or doing yoga, move your body to help release the tension that has built up from these situations. Remember, you can't control what's happening to you, but you can control how you handle yourself.

- Talk to a confidant. Teaching is **hard**. It's hard enough trying to manage thirty or forty different personalities in one classroom, let alone mastering curriculum design and bringing all your students to academic proficiency. Add on the politics and bureaucracy, and you have a formula for pressure overload. Lucky for you, you have a *ton* of colleagues going through the exact same things. Lean on them.

- Finally, and most importantly, never lose sight of how important you and your presence are to the students you get to work with each day. Your strong, consistent presence is oftentimes the stability and strength they need to grow and empower them to be their best.

Moral of this story: If something seems wrong or unfair or confusing or inappropriate, chances are there are politics at play and systems out of your control. Fed up with it? Join me in the movement to massively change the system and bring in and retain individuals who truly inspire and empower future leaders of America. We got this.

 Have a specific educationally-related political question, comment or concern? I want to know what you think!

Head to **withheartproject.com/contact** so we can chat!

Chapter 7: Own It

In my third year of teaching, I won a "Teacher of the Year" award. I'm not saying this was the standard "Teacher of the Year" award where it's basically someone who has been teaching for forty years in the same district and finally gets recognized for their teaching after sticking it out for so long. This "Teacher of the Year" award came directly from my students. The local Wal-Mart had a booth where students could elect teachers they felt would be deserving of this award, and several of my AVID students elected me. The day before was a regular school day with staff meeting at the end. I was crazy sick. I had caught the latest version of whatever virus was going around at our school. As such, I had already determined that I would not be going to school the following day. I needed a break. That day, my husband told me I couldn't stay home and that I *had to* go to school. This was incredibly unlike him. I was really stubborn and gave him the third degree about why I couldn't stay home. He said something to the effect of, "Look, something is going to happen at your staff meeting that you need to be there for. Just go." So, I sucked up all the energy I had and went.

Toward the end of the meeting, an announcement was made that a local business woman would be recognizing a special teacher. To be perfectly honest, I was only halfway paying attention given how awful I felt, so when I heard my name called, I didn't know what I was being recognized for and was totally embarrassed. I went up on stage as my teaching colleagues applauded, and the businesswoman thankfully explained what I was being recognized for; she handed me an award along with photocopies of all the students' submissions explaining why they elected me. I remember walking back to my table and one of my math teacher colleagues said to me, "Congratulations, and dang it's so early in your career!" I was so touched and humbled (and still not feeling great, so I went home to pass out and sleep off my cold).

Winning this award really helped boost my confidence in terms of my teaching ability, since I started my teaching career on an internship and genuinely was learning as I went. I would often check in with the website ratemyteachers.com to see what other kinds of ratings I was getting, and I how seemed to rate compared to my colleagues. Throughout the following decade in the classroom I was recognized for a variety of things: department chair/lead, curriculum writer/developer, pathway writer/developer, was selected among 100 other teachers in the entire country to work on high-quality math lessons with the educational organization LearnZillion, and other things attesting to my teaching ability.

I also passed out evaluations each school year so my students could give me direct feedback about how I was doing serving them as their teacher. Whenever I got less than positive feedback, I would look into it, really **hear it**, and make changes to be better. Overall, I had a great career, which was why a situation that erupted in my classroom during my 13th year in education shattered me.

I had spent all school year working with my students on the importance of being respectful to each other, boosting communication skills, social-emotional skills, holding community building circles, etc. Other teachers on campus would come to my classroom just to observe the "amazing culture" I had created. This year in particular, I made a huge effort to infuse my math class with restorative practices, positive behavior interventions and supports (PBIS), and explicit social-emotional lessons. I had spent one day per week *not* teaching math, and *explicitly* teaching social-emotional lessons, with topics including empathy, anger management, compassion, building relationships, assertiveness vs. aggressiveness, violence, communication skills, rewriting negative thinking, mindfulness, and many more topics. By the end of the school year, my kids were helping to *prevent* fights from happening on campus. So, with about a month left in the school year, on our typical "Wellness

Wednesday," I didn't hesitate when one of my students, Kelly, asked if we could do a circle. I said we could, and she asked if she could pick the topic. I obliged; we had built such a great culture in my classroom that I was convinced nothing could go wrong. I was wrong.

The topic she chose was, "Name one thing you like about this class, and one thing that frustrates you about this class." Per our usual routine, I went first to model appropriate responses. I passed the talking piece to a student to the right of me, and the talking piece made its way around the circle. It came to four girls who happened to be good friends and were generally louder and more jovial than other students in the class. They each took turns telling the rest of us what they liked, and what frustrated them, and we moved around the circle. The talking piece got to a very quiet, mild-mannered student, and the dialogue went something like this:

Jane said, "The thing I like about this class is that Miss Miller is really helpful and really cares about us. The thing I don't like is..." (her tone changed from mild-mannered and happy to irritated and angry in a split second) "...I'm sick of *certain people* always being so loud and not doing what Miss Miller asks us to do."

Kelly responded, "I feel like you're talking about us, are you talking about us?"

Jane said, "Yes, I'm talking about you."

One of Kelly's friends, Emma, said, "I understand what you're saying, but you don't have to say it so rude."

Jane replied, "I'm sorry, but I can feel however I want."

Kelly got heated and stood up, punching her fists together and saying, "Now, I just want to beat someone up."

At this point, I was still calm and collected. I stood up and said, "Okay, ladies, head outside, take a deep breath and cool down." Kelly and Emma made their way toward the door but stopped *next to* the door and turned around to face the wall, still mumbling under their breath. I again asked them to go outside, but they wouldn't. Their continued sputtering under their breath made Jane continue to engage with them, saying, "I'm sorry, but I'm entitled to feel what I feel." This continued to make Kelly and Emma mad, so to interrupt the negative conversation, I physically put my body in between the girls. They were about ten feet apart and not physically threatening to one another, but I felt better being physically between them. The two girls continued mumbling and grumbling under their breath, so I yelled to them, "Ladies, just shut the f@#$ up!"

I continued, "Look, the bottom line is that everyone in here is entitled to feel what they feel and communicate what they feel to one another as long as it's in a respectful way. Jane has every right to feel what she feels, and she was trying to communicate that to you in a respectful way. The problem is that *some people* can dish out their feelings and insults to other people but can't take it when someone has something less than pleasant to say about them. Ladies, you need to grow the f@#$ up!"

They stormed out of my classroom and the rest of the class was sitting there looking at me waiting to see what I was going to do next. In my thirteen years in education I've *never* yelled and spoken to any of my students with that level of anger. I apologized to the class and said that that was an unprofessional way of acting and that neither the two girls nor the class deserved that. I felt *awful*. I felt like a hypocrite. I spent the entire school year building an environment that was the exact opposite of what I just modeled. I spent the entire school year talking about how to manage yourself when things get heated. But I erupted and probably destroyed the relationships with those students in a matter of seconds. (More on why in just a bit.)

They ended up going to the office and writing a witness statement to report my conduct. This was the last class of the day, so when class was over, I went straight to my department chair and told him what happened. I was the third math teacher that I was *sure* was going to be put on paid administrative leave for the rest of the year for my conduct. One teacher allegedly put his hands on a student, and another teacher allegedly told an African-American student to sweep the entire classroom in front of the class. The overall climate of our school was absolutely contentious. The department chair and I went straight to the principal's office and told her what happened. I apologized profusely for my conduct and explained the details of the situation to her. She said she would have to call the Director of Human Resources to talk about next steps and would get back to me later that day. I went home sad and drained.

When she called me later that day, she told me that I would not be put on paid administrative leave, but that I would be getting a letter of reprimand. She told me that I should come to work tomorrow and that the school needed me. At that point, I recognized I needed a day to rest and recharge. I took one. I happened to be at a training the following week and had the opportunity to use those days to recharge before coming back to the environment I worked so hard to build. During that time, I called both parents of the students I yelled at and told them what I did and how sorry I was. One of the parents actually said, "I'm not mad at you. I actually wanted to thank you for what you said because now it's not just me telling her she needs to grow up." It was a truly confusing experience, but not surprising, because not only had I spent the entire school year building relationships and community within my classroom, I also built relationships with certain students' parents, one of which was Kelly's mom. She knew I had her daughter's best interest at heart, and with that, she was on my side even though I told her daughter to, "shut the f@#$ up" and "grow the f@#$ up."

When I returned, Kelly and Emma refused to look at me. I explained to the class that we needed to circle back around and get some clarity and closure about the volatile interaction. We left the situation on a very negative note, and it wouldn't be right to just press on as if nothing happened. I had been formally trained at that point to facilitate restorative conferences, so I started in with the questions I was trained to use in situations of conflict. Kelly talked about what happened from her perspective. She explained that at home, she has a really loud family - that's just how they are. Jane explained that she has a really quiet family - that's just how *they* are. I used that moment to point out that even though they are different and have different backgrounds, neither of them is right or wrong, they're just different. They both apologized to each other and the overall tension in class was lifted. Everyone was laughing with each other by the end of the class period.

I again apologized for my part and explained to my students exactly what happened from my perspective. I praised Kelly and Emma in front of the class for reporting my unprofessional behavior to the office. This was something I had been trying to teach them all year. I explained to them that my consequence was getting a letter of reprimand, and that we would no longer be able to do "Wellness Wednesdays" as part of the consequence. I explained that I felt like a hypocrite because I did everything I told my students *not* to do in that moment. I hadn't taken the time needed to take care of myself. I was completely unbalanced in terms of self-care. I used that time to stress to my students the importance of checking in with yourself and taking care of yourself, so you don't erupt in tense situations. The year ended well, with both Kelly and Emma expressing to me that they were sad I wouldn't be there next year. Kelly even said, "Miss Miller, I don't know what we're going do to without you next year."

So, I told two students to "shut the f@#$ up" and "grow the f@#$ up," and they were sad I was leaving, and their parents thanked me for talking to them this way. I can 100 percent tell you that there are other teachers who say similar things to their students, and their students hate them, refuse to do work for them, and try to get their parents to get the teachers fired. So, what is the difference? Why did my behavior end positively when other teachers' similar interactions made their classroom environments negative or even worse than they were before the interactions? The answer is simple — **ownership**.

Behavior Management Style

Students misbehave all the time. Students push your buttons all the time. That's what they do; it's part of their job description. That's part of growing up and learning about yourself and your identity. The difference between a teacher who's hated and a teacher who's loved is that the teacher who's loved can "own their sh@#." They can look themselves in the mirror and really see how their own actions and words may be negatively affecting their students and their classroom environments. They have the ability to see how their words and actions affect people. They are willing to modify their words and actions to better meet the needs of their students. It is not for the faint of heart to be a teacher who is truly loved. There is a *major* difference between teachers who are willing to change themselves for the sake of meeting their students' needs and teachers who consistently find reasons to support their ineffective teaching and behavior management practices.

Let's dive into this a bit. We're going to take some time to really look at ourselves as individuals and see where we are in terms of our teaching styles, triggers, and implicit biases. Let's start with behavior management styles. The figure below shows the social discipline window from restorative practices, adapted by the International Institute for Restorative Practices (IIRP). This window shows four types of behavior management styles: permissive, punitive, neglectful, and restorative. Read the descriptions below and circle the behavior management style that sounds most like you.

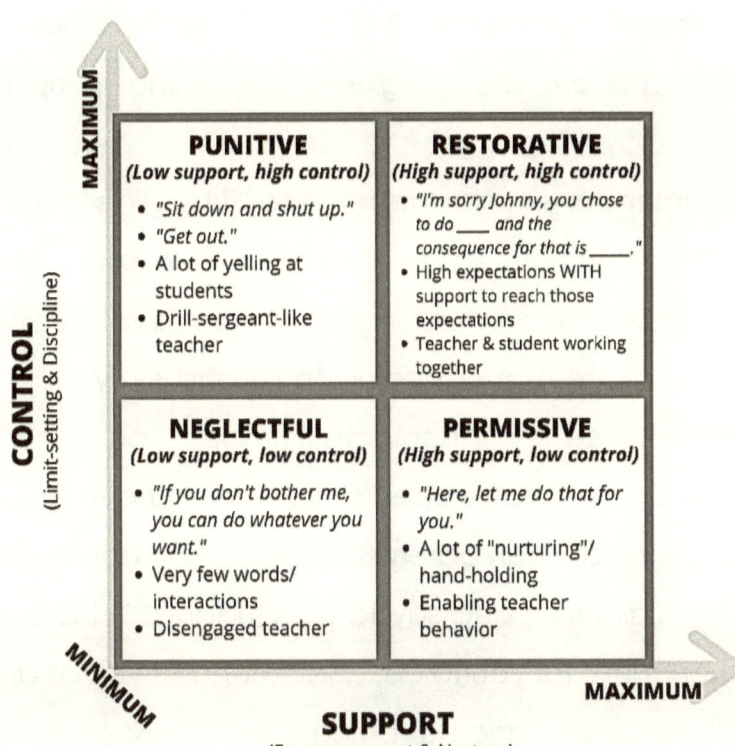

Many teachers fall in the permissive, punitive or neglectful categories, simply because it's easier. When students misbehave, it's much easier to send them outside or suspend them than it is to have an at-length discussion about why their behavior was inappropriate and teach them how to correct it for next time. As teachers, we equate doling out a consequence to the student learning his/her lesson. Unfortunately, it doesn't quite work that way. It's also much easier to, when students are struggling with something, jump in and say, "Here let me do that for you." The problem is when students are struggling, that's often the time when they learn the most. The struggle itself is where massive amounts of learning take place. I'm not going to lie and tell you I've never been in "neglectful" mode, because I have. There have been days where I have been so sick that I can barely stand, and I will come into my classroom and say to my students, "I have a worksheet for you to do, and as long as you all are sitting quietly and not being loud, disrespectful or inappropriate, you can do whatever school-appropriate activity you want." These are self-care days and happen at least a few times each school year. But these days are not my overall teaching style.

When you as a teacher take time to work with students, to teach them how to be self-sustained, to empower them and actually give lessons about how to not make the same mistake again, you are truly being restorative and kids will feel that in their bones. If you are a parent, think about when your own children make a mistake. Do you just send them to their room and not have a follow-up discussion about why what they did was not okay? Probably not. You probably sent them to their room to take a break, then looped back around and had a discussion with them about why their behavior wasn't appropriate, and how to not do the same thing again. **That** is where the true learning takes place. I know it's one thing to be parenting your own child, or two or three children, and a totally different thing to be a teacher of 35-150 students. But if you can try this approach with just one or two students, it may make all the difference.

Implicit Bias

We're going to switch gears now to another topic: implicit bias. What exactly is implicit bias? If I look it up in the dictionary, I get the following definition:

> "the attitudes or stereotypes that affect our understanding, actions, and decisions in an unconscious manner. These biases, which

encompass both favorable and unfavorable assessments, are activated involuntarily and without an individual's awareness or intentional control. Residing deep in the subconscious, these biases are different from known biases that individuals may choose to conceal for the purposes of social and/or political correctness."

During the latter years of my experience in education, we started receiving implicit bias training, and this is one that if taken seriously and truly looked at with an open mind and open heart, can be transformative to your teaching practice.

I'd like to start by asking you to take the Implicit Bias Test, which can be found at www.implicit.harvard.edu. There are several tests to choose from; a good one to start with is the "Race IAT" test. The test takes around ten to fifteen minutes and will ask you a series of questions to identify your implicit biases. At the end of the test, you should end up with a result that says something like this:

"Your data suggest a strong automatic preference for: Dark-Skinned People Over Light-Skinned People."

Write down your implicit bias here:
(Seriously, pause now and take this test before moving on.)

_____

_____

I'd like to be clear here and make sure you don't beat yourself up over this. Some people get their results and automatically think, *Oh, my gosh, I must be racist! What's wrong with me?* But that's not necessarily the case. Ultimately, it's our experiences that have shaped us to be the person we are today. You ended up with your preference based on your experiences, but that doesn't mean you are stuck that way forever. If you want to make that change to yourself, you have to begin by identifying what is living in the subconscious parts of your mind. Only by shedding light on what's going on in your brain will you be able to be more mindful of the way you're showing up for your students, and therefore have the ability to change to better meet their needs. We can't change what we don't know.

My thirteenth year in education was truly transformative in many ways. Not only was there the situation where I cursed at my students in a moment of anger, but there also was another situation — one that really made me stop and think about how to handle it, even though I had been handling it in a very specific, detailed, process-oriented way in the prior twelve years. As teachers we are given very specific training if we suspect a student is involved in some type of child abuse. Any suspicion? Call Child Protective Services (CPS) and report it. Shortly after my heated interaction with Kelly and Emma, things had gotten back to normal in my class. So, when Emma showed up to school and told me her entire body hurt, I couldn't ignore it. I pulled her outside and asked her to tell me what was going on. After a lot of back and forth, she told me that her stepfather hit her the night before. My response was something to the effect of, "Oh, absolutely not, that's not okay."

Kelly was right by her side to assure me that "that's just how black people discipline their kids."

She even went so far as to ask any African-American student that she saw walk by, "Hey, when you get in trouble at home, what do your parents do?" And one by one, every student said they get "whoopins." At the beginning of my teaching career, I wouldn't have even considered not calling CPS, but I figured maybe this was truly a cultural difference. I told my students that I was a mandated reported and had to call CPS in situations like this, but that I'd do some research before calling (at which point they made fun of me, but I digress). I went to my principal and told her the situation, and that ten years ago I wouldn't even hesitate and would call no matter what, but that I was trying to be more culturally aware in my teaching practice. She suggested to call CPS and explain the situation and that I wasn't sure whether it was report-worthy or not. So, I did.

I called and explained to the CPS worker the situation, and the argument that, "that's just how black people discipline their kids," and at that point the CPS worker asked for the name of the student. I gave it to her, and it turned out that there was a long history with this particular student where abuse was evident from a very early age. I made the report and CPS followed up with a visit. Emma ended up being extremely angry with me, but she still confided in me with other things because deep down, she knew I had her best interest at heart.

There are a couple of lessons to learn here:

- Never question abuse if it's under the guise of "cultural discipline."

- Use cultural differences as a method for teaching acceptance and compassion. After my explosion and telling Kelly and Emma to "shut the f@#$ up" and "grow the f@#$ up," I used that time to let Jane and Kelly see how they came from two very different cultures, and that was what was causing the tension. I continued to explain that different doesn't mean bad, it just means that we *all* need to practice increased mindfulness and compassion for people who are different from us. Only then will we be able to move forward and create positive, respectful, compassionate, and nurturing learning environments where we truly care about each other as people.

Identify Your Triggers

We all get triggered from past events and circumstances that led up to this moment here and now. We're going to take some time to determine what *our* triggers are and ways we can stop them from causing full-on blowouts like mine with Kelly and Emma. Take a few minutes and answer the questions below, which were adapted from *Fostering Resilient Learners*:

IDEAL STUDENT BEHAVIORS AND ATTRIBUTES Your idea of how students **should** behave.	UNFAVORABLE STUDENT BEHAVIORS, ATTRIBUTES The ways students behave that **irritate** you.

How do you typically respond to your least favorite student behaviors, attributes, or triggers?

_____

_____

_____

What are some more appropriate ways you could respond to your triggers?

_____

_____

_____

One of the most important things to note is that your students can very easily spot your triggers, even if you haven't shared them explicitly. Students are very observant and can trigger you for a variety of reasons. Remember the Teacher Gone game?

- It's a fun game for them.
- It allows them to avoid doing the task at hand.
- It's hard-wired into them to see what they can get away with.
- It allows them to not be held accountable for their actions.
- If you get so flustered and caught up in their arguments during your trigger moment, the focus shifts from their inappropriate behavior to your reaction.

Many teachers maintain a level of curtness or drill-sergeant like behavior management styles in their classroom to make sure order is maintained. This is certainly a way of managing behavior, but one that really should be looked into. In this way, their reactions aren't triggered as much because they've conditioned themselves to ignore behavior so much and have quick responses to shut bad behavior down. A word of caution, though: putting up a wall and managing your classroom without letting students in and really allowing them to see you as a person stands not only between you loving and hating your job, but also between staying sane in your classroom while also keeping students in your classroom.

So, the next time you are triggered, take a deep breath, then try something new...

 Have a specific "owning it" question, comment or concern? I want to know what you think!

Head to **withheartproject.com/contact** so we can chat!

Chapter 8: Why Not Try Something New?

If you've been in education long enough, you know that many movements and shifts have occurred in instruction and practice that we've had to work through and keep up with. It's exhausting. When I entered the education field, it was all about the STAR Test. In these tests, all students were tested on all subjects they took in high school every year. It consisted of a long laundry list of multiple-choice questions that students had to answer to show they were proficient in the subject at hand. During my fifth year of teaching, I was challenged to double the proficiency on the Algebra 2 STAR Test. At this point in my career, I was working at a very high-performing, high achieving, "little-to-no-extra problems," affluent school. There should have been no reason I couldn't double the proficiency from 19 percent to 38 percent.

So, I did. I busted my tail and implemented a "mandatory tutoring" program in which any student who received a "D" or an "F" on a test was required to come in to my classroom during lunch for tutoring. I allowed my stronger students to help students who were struggling for extra credit, and over time it was like a well-oiled machine. I was sure my students were going to kill it on the test that year. In fact, I was so sure that this would work, that I was ready to celebrate a 75 percent proficiency rate when it was all said and done. Except I didn't get to celebrate the 75 percent proficiency rate. I did, however, get to celebrate that the Algebra 2 STAR Test proficiency rate went up to 41 percent. I met the goal initially given to me by my supervisor. But I was perplexed.

I didn't understand how I could implement these programs to put more accountability on the students, essentially teach directly to the test (from the released practice test questions), spend each and every day teaching Algebra 2 from bell to bell *and* give extra help during lunch, before school and after school, and still have only 41 percent of students achieving proficiency. It didn't make sense.

At the same time I was busting my tail to increase the Algebra 2 proficiency, I began to get extremely frustrated. I understood that we were a charter school and that funding was allocated largely based on students' test performance, yada, yada, yada. But I hated it. The bottom line was that my students were *people*. They weren't robots there to just absorb information and regurgitate it on a test to make their teacher look good. I'm sure they felt that pressure to increase the overall proficiency, even though I never told them outright that's what was happening. But when given that kind of goal, as a problem-solver, I went into problem-solving mode. I thought to myself, "Well, if I want to double the proficiency, I can't waste any time. I *must* teach math from the minute they enter the door to the minute they leave and cram it down their throats in their after-school time as well. After all, practice makes perfect, right?" So, my goal was well-intentioned, but it completely ignored the fact that these were actual human beings with their own personalities and sets of issues, and that they had a higher purpose in life than just spitting out Algebra 2 facts on a standardized test. I was bothered. Really bothered.

I spent the next few years at that school, but then ultimately realized that I missed working in an environment where it was okay to acknowledge that kids are people and have a wide range of needs, and that it was okay if they didn't score proficient on one standardized test. I went back to my second home for nearly the remainder of my teaching career: a high-poverty high school.

What made high-poverty schools such a right fit for me was that kids needed me in a way that they couldn't get filled anywhere else. They needed me to guide them to be whole, beautiful, imperfect human beings, yes academically, but *also* socially and emotionally. Many students over the years have confided in me with their struggles in life, and I enjoyed that aspect of teaching. Many teachers in high schools truly hate this aspect of teaching and want to be able to just come to school, teach their academic subject, have all their students be proficient, go home and call it a day. But I wasn't like that. I'm still not like that. I believe it's *crucial* to teach the whole child.

So, when I began my thirteenth year of teaching at a high-poverty middle school, I threw out everything I had done previously and ran my classroom in an entirely different way. I put relationships at the forefront. I put my students as people at the forefront, and math was secondary to all the other aspects of what made these students human. I approached my administration and threw out an idea: I wanted to start teaching explicit social-emotional lessons once per week, on Wednesdays, and I would call it "Wellness Wednesdays." This is something that had never been done before in an academic class, and as such, was truly out-of-the-box. My principal asked me to draft a letter to parents and let her look at it before making any definitive decisions. Here is what part of the letter looked like:

Dear Parents/Guardians,

This year ▮▮▮▮ School is placing a large focus on social-emotional wellness. In an effort to embrace the whole child, including his/her overall wellness, students in my Math 2 course will be spending some time on Wednesdays taking a break from the academic curriculum as outlined in the syllabus you signed earlier this year, and focusing on other things to help build students' confidence, resilience, perseverance, etc. As such, I wanted to give you a brief outline of the topics we will be covering; the grid below gives a description of each topic, lesson and activity; please take a minute to review these topics, then sign the back page acknowledging you've received this information.

Topic/Lesson	Lesson Description	Activity Description
Empathy	We discuss what empathy is, as well as various steps kids can take to build empathy skills.	Students identify various people in their lives to which they could show empathy, and discuss the steps they could take to show additional empathy to each person.
Labels & Stereotypes	We identify and discuss various labels/stereotypes that exist at our school, then discuss why these labels/stereotypes might be hurtful. Students take a stereotype survey, then look at the typical progression that leads from labels/stereotypes to anger, hate and violence.	Students identify various groups on campus, then discuss ways in which stereotypes/labels can be hurtful.
If you Really Knew Me	Students are paired up with another student, and asked several lighthearted questions to identify similarities using a Venn diagram. Students then make a "Words Have Power" goal, identifying various things they say/do that could be hurtful to another student, then are challenged to live up to their goal.	Students share information they would like to share with the teacher in an effort to get to know each other better. They also discuss things they've been through, as well as things they haven't been through but have empathy for.
Life Stories	We discuss the idea of getting to know ourselves better in an effort to celebrate each student's individuality and life experiences, use their experiences to build a solid foundation to continue to grow on.	Students create a poster to represent various aspects of their life to this point in time in an effort to celebrate each student's individuality and life experiences, and use this as a solid foundation to continue to grow on.
Self Esteem & Positive Self-Talk	Students are given examples of various teenagers around the world who have broken social norms by helping build other kids' self-esteems. We go on to discuss the power of positive self-talk and how self-esteem is shaped; the danger of social media with regards to self-esteem, give specific examples in life where positive self-talk	Students take a "self-esteem survey," then rewrite common negative sentences/thoughts into more positive ones. This is the second step in helping students see themselves in a more positive light, and shoring up their foundation as they move

I emailed the letter to my principal and continued with my duties. She responded to my email after a few days and said, "This is amazing. I think the whole school should be doing this. Go ahead and send out your letter and start!"

The first Wednesday, I introduced the idea to my students, and assumed we'd be working through the lesson in about 20 to 30 minutes, then would get back to math for the remainder of the class period (we were on block schedule, so that would've left roughly an hour for math concepts). The first lesson was on empathy. It ended up taking the entire ninety-minute block of time. Kids were engaged and asking questions and wanting to learn more and more. So, we pressed on throughout the year working through various Social-Emotional Wellness (SEL) concepts one day per week.

Being the data nerd that I am, I wanted to see what kind of impact this had on my students' academic growth, so I approached my fellow seventh grade math teachers and asked if they would allow me to compare my students' diagnostic assessment scores at the halfway point of the year to theirs. After all, I was skipping one entire school day each week to focus on explicit SEL. I had a hunch this would help them in the long run, but I needed numbers to prove it.

Here's where we landed: When it was all said and done, I taught math 67 percent of the school days in the first semester, and explicit SEL 33 percent of the school days. The other three math teachers taught math approximately 98 percent of school days, and explicit SEL two percent of school days. The days they taught SEL were at the beginning of the school year in the traditional "get to know you" relationship building activities. *My students outperformed all other seventh grade math students on campus.* Now this doesn't mean that every student in my classes did better than every student in other classes, but on average, my students grew mathematically more than all other Math 2 students (detailed data is available on www.WithHeartProject.com).

I was right. Kids **needed** to feel like they were being led by someone who cared about them as individuals, not just as data to support school statistics. My students knew I cared about them very deeply, and that my primary focus was on them as people — all of them, not just their math ability. It was a complete paradigm shift, and one that in my professional opinion, is going to be the next large movement in education going forward. But when it comes right down to it, are we ready? Are we really, truly ready as a society to make this huge shift?

The Director of Social-Emotional Learning for this district wasn't so sure. She came to observe one of my Wellness Wednesday lessons, and had some feedback for me, which was both positive and constructive. The lesson she happened to come observe was around the topic of violence. We had spent ten to twelve weeks working through concepts on Wednesdays that were more positive, but at some point, we had to get to the realities many kids are facing nowadays. So, one of her main questions for me was whether students were triggered by the examples of violence in my presentation.

"I'm glad you asked. Yes, they were. I had two students who were triggered by the lesson," I said.

"Really...hmmm..." She was tentative in her response.

I replied, "One student in the class ran outside crying during the lesson and sat outside while we finished. Once I had the class working on the activity that went with the lesson, I went out and asked her what was going on. She eventually confided in me that her father was raping her. The other student, in the second class I taught, had a similar reaction. She also ran outside crying, and I eventually made my way to her to see what was going on. She said she had plans to kill herself after school that day."

The director looked concerned, but I continued.

I said, "Honestly, I'd rather have them be triggered in a safe environment and have them be able to confide in me things that are going on at home, so that I can get them the help they need, than for them to not be triggered and still continue suffering unnecessarily. My lessons essentially allowed me to intervene and make it so the one student is no longer getting raped at home. The other student now has the support she needs to safely handle her depression and suicidal thoughts."

The director nodded along as I explained, and eventually said, "Ya know, Kristen. I think you're out here (motioning with her hand) and the district is back here (motioning with her other hand). You are where we want to be, but there are about a million tiny steps between that the district needs to take to get us there."

She invited me to be part of the SEL Guiding Coalition for the district, and I helped design the mission and vision statements, as well as an action plan for rolling out SEL district-wide in years to come. So that's where we are. A massive, massive paradigm shift. We're no longer teaching in an environment where students *need* a teacher to teach them academic content. If a student truly wants to learn how to solve equations for x, they can look it up on YouTube or Khan Academy. Students don't *need* their teachers for academic content. What they *do need* from their teachers is connection — human connection and guidance about how to deal with all the pressures and challenges they face in their personal worlds, hindering them from being successful academically. So, this is my challenge to you. I've already proven that less is more when it comes to teaching academic content, but are you ready and willing to shift your instruction to focus on your students as people rather than academic content regurgitators? Here are some strategies that may help you get started.

Less is More

As teachers we often believe we must teach bell to bell. If an administrator walks into our classroom and sees us doing something other than teaching the academic content at hand, we're afraid we'll be in trouble. But here's the thing. When kids *feel* that you don't care about them as human beings, they shut off. If you can get kids to see you truly care about them as individuals, not just as test scores, they will *want* to work. It seems counterintuitive, but I promise it's true. I'd like to introduce you to the 75/25 rule; 75 percent academic content teaching, 25 percent social-emotional, restorative practice, trauma-informed practice teaching. So, if you have fifty-five minutes of instruction for each class period, maybe try fifteen minutes of non-academic content and the remaining forty minutes on academic content. Okay, it sounds great, but what exactly are we supposed to do with these kids during the fifteen minutes of non-academic content time? I'm glad you asked.

Learn Your Students' Names!

Within the first two weeks, you should have all your students' names memorized. There's nothing else to explain here. If you can't take the time to get to know your students' names, why in the world did you choose this as your profession?

Circles

Circles make the world go 'round. This is a highly effective strategy to use with students and one that even the most disengaged, unmotivated student will end up looking forward to. What exactly is a circle? It is where you literally arrange the chairs in a giant circle, one for each student, and sit there and stare at each other while doing an activity. What are some activities you could do in a circle?

Community-building circles are a great way to allow students to get to know each other, and for you as the teacher to get to know your students. There is a protocol for circles: Only one person speaks at a time and the person who speaks is the one with the talking piece. A talking piece can be anything; I used a small stuffed animal or a small earth-globe squishy ball. It could be as simple as a rock or a talking stick. The idea here is to teach kids the importance of respectful listening when someone is speaking and practicing patience. There are a variety of questions you can ask, but generally speaking, these should be **non-academic** questions that literally focus on getting to know your students as people. Sample questions/prompts could include:

- My favorite sport is _____ because _____.

- When I grow up, I want to be a _____ because _____.

- My favorite hobby is _____.

- I am happy when _____.

- If I were a famous actor or actress, I would be _____.

- When I graduate from high school, I want to _____.
- I can't wait until _____.
- Share a happy childhood memory.
- If you could be a superhero, what super powers would you choose and why?
- What gives you hope?
- What demonstrates respect?
- Share a time when you were outside of your comfort zone.

There are a ton of resources out there to give you ideas for topics during community-building circles. Just Google "community building circle prompts" or visit www.WithHeartProject.com for additional ideas.

Mindfulness

With all the technology and social media and video games floating around these days, kids seem to be largely incapable of just *"being."* Not only is the increase in technology impairing physical health in a variety of ways, it's impairing mental health for a larger number of adolescents than ever before. Getting kids away from their phones, computers, tablets, video games, televisions, etc., is going to be *huge* in helping them be more well-rounded, healthier individuals. Some activities to practice mindfulness:

Mindful Breathing. Okay, so I know what you're thinking: *There is no way my students will be willing to practice breathing in my class. We're talking about middle and high school aged kids. They'll all just look at me like I'm crazy if I try to bring in some breathing activities.* This is where it is *crucial* for you to explain the neuroscience of what's going on in their brains, so here it is:

TRIGGERED BRAIN
(Flipped Lid)

NON-TRIGGERED BRAIN

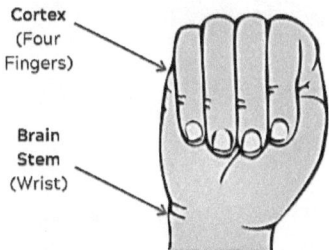

Cortex
(Four
Fingers)

Brain
Stem
(Wrist)

Amygdala
(Thumb)

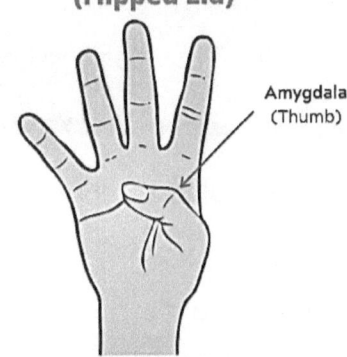

- Student is relaxed and/or calm
- Student can think rationally
- Student can focus on academic subjects
- Cortex in the driver's seat (responsible for higher-level thinking)

- Student is in fight-flight-freeze mode
- Student can NOT think rationally
- Student can NOT focus on academic subjects
- Student is highly reactive and emotional
- Amygdala in the driver's seat (responsible for sounding the alarm)

Details of how to actually DO mindful breathing with your students can be found on www.WithHeartProject.com.

Mindful Walking. I had my students participate in mindful walking during one of our Wellness Wednesdays. I had a grassy area outside of my classroom, and instructed students to walk very, very slowly around the grassy area not talking, not interacting with anyone except for themselves. They were to focus on feeling their shoes go from heel to toe on the ground, how the air felt hitting their face, how the wind rustled their hair, and any smells they could smell. I had them do this for seven minutes. They loved it so much that they asked to do it again and again throughout the year.

Mindful Listening. I have a Tibetan singing bowl in my classroom, and I would challenge students to listen very closely when I rang the bowl. If you've never heard how a Tibetan singing bowl sounds, you hit it once, and the ring carries on for a very long time. So, I'd have students all get very quiet and I would go in the center of the room and hit the bowl. I'd ask for students to raise their hands when they no longer heard the ring. They got a kick out of it and wanted to do it again and again.

Wellness Writing

This doesn't have to be a huge or invasive thing, and I'm not talking about academic writing. I'm talking about journaling. When taking teaching credential classes, we learn a lot about the importance of structure and having a warm up on the board as students come in to help get them into a routine. Having a warm up also gives you time as the teacher to take attendance and take care of any opening activities you may need to do before diving into academic content. Each day my students came into class, there were two warm ups on the board: a "Wellness warm up" and a "Math warm up." A few examples of a wellness warm up might look something like this:

- **Today I feel _____ because _____.** I would have emojis on the board so they could give me a visual representation of how they were feeling, and I would encourage them to write about whatever was making them feel this way.

- **Before I got to class today, _____, and that made me feel _____.** Again, have them elaborate on the situations that came up before entering your class.

Give students around five to ten minutes to write their warm up, then ask for volunteers to share out loud what they wrote.

Greetings

I used to hear so many educators talk about greeting students at the door, and I never believed it would be beneficial — until I did it. If you're anything like me, you don't do things just because someone else tells you to do them; you do them because you believe they're worthwhile and will make a difference. The reason I chose to stand outside of my door and greet students every day was not because my principal told me to, but because it allowed me to get to know my students better. You can tell a lot about your students simply by looking at their facial expressions, their body language, tone, etc. If you see that Johnny is normally happy and cheerful when he shakes your hand, and suddenly he's sad and withdrawn, that would be a good time to have him step aside before he comes in and ask him if he's doing okay. Remember, two questions:

- Are you okay?
- What's going on?

Then carry on a conversation with him like he's a real human being that you want to make sure is okay.

Flood Your Students with Positivity

Recent research shows that adults are highly unsatisfied in their careers and people are quitting their jobs like never before. In his book *Lead from the Heart*, Mark Crowley describes a research study that found that the things people most cared about were that their boss exhibit appreciation and recognition, that they feel like they're making a valuable contribution, and that they're able to trust their boss.

Your classroom should be looked at as a mini workplace where you are the boss. Think about how good it feels when you get recognized for working so hard and/or putting in a lot of effort. Your students are just the same. Even though they're in middle and high school, they still want and need positive encouragement and recognition.

Snowball Fight

If you sense the energy in the room is off, it might be time for a snowball fight. What exactly does this mean? Have each student take out a piece of paper and write down whatever is on their mind, especially if it's frustrating. Encourage them to write with enthusiasm. Sometimes this means writing so hard that the paper rips, but this is a way for them to get out their frustration in a relatively harmless way. Once they finish, they're going to crumple up their paper into a ball and stand up in a circle around the classroom (or outside if it's easier). On the count of three, they can start throwing their papers at each other. Before they start throwing, give them a hand signal to show them when they need to stop throwing. These kinds of activities are *great* as long as you set firm boundaries before you begin. I would tell them they could throw the papers at each other until they saw my hand go up. That was the signal to stop. The other hard and fast rule is absolutely no physical contact with each other in any way shape or form. They can throw papers at each other, but not body parts. Give them a couple of minutes to continue throwing the papers at each other, then when all done, have them work together to find all the snowballs and put them in the recycling bin.

Share Your Own Stories

In working with a teacher who was having difficulty with a student, I suggested that this teacher share his personal story of his background with the student. He scoffed at the idea and said point blank, "Nope. Can't do that. I can't lose my street cred." I thought to myself, *What the heck? Street cred? This is a school, dude. Get over yourself.* If you're only focusing on academic content and emitting this aura that you're a perfect human being who never makes mistakes, you will not be able to connect with your kids. These kids are *craving* connection. Remember, they can look up how to graph linear inequalities on YouTube. They can't make a human connection on YouTube. This is quite possibly one of the most powerful things you can do for your students. I know what you're thinking: *If I share my own stories and mistakes with them, they'll lose all respect for me and my class will quickly get out of control.* But in reality, it's the opposite. The more you're willing to let them see you as a human being, the more they will *want* to work for you and get in control and on task if and when you ask. I know it can be scary, but please give it a try.

Be Vulnerable

This goes hand in hand with sharing your own stories. But with any and all of the activities I list above, the more you participate and show your students the human side of you, the more they will respect you and the more effort they will put into working for you. The idea of being vulnerable in front of your students is scary, but it's so worth it if you can get out of your own way.

If You're Feeling Adventurous

Explicit social-emotional lessons are a great way to really connect with your students. It is *in* these lessons that you will truly be able to forge strong relationships and show them the human side of you. You saw some examples of lessons I did with my students, which they loved. There are many high-quality lessons available on www.TeachersPayTeachers.com. I am in the process of creating videos and subsequent activities to go with each lesson for other teachers to use in the future. Please check www.WithHeartProject.com for lessons as they become available.

When it's all said and done, it's up to you. You can continue teaching in the same way you've been teaching and continue to have little to no connection with your students and the same results year after year. Or you can try something new and create strong, meaningful relationships with your students — and potentially reignite your love of teaching. I challenge you to try at least one of the strategies listed above in one of your classes once per week.

Are you up to the challenge?

 Have a specific question, comment or concern about trying new things? I want to know what you think!

Head to **withheartproject.com/contact** so we can chat!

Chapter 9: Empower Yourself

Before I was a teacher, I was an engineer. The way that job looked from day to day was *boring* compared to a teacher's day. Here's what my day looked like:

- 8 a.m. to 12 p.m.: Work on structural drawings and calculation packages, call other engineers to verify various components of the work, and work with drafters.
- 12 p.m. to 1 p.m.: Lunch.
- 1 p.m. to 4 p.m.: Work on structural drawings and calculation packages, call other engineers to verify various components of the work, and work with drafters.

There were days where I literally almost fell asleep at my drafting table because I didn't have a lot to do and/or had very little stimulation other than numbers and computers. In fact, some days I would show up to work and complete the task at hand, and still have three hours left in the day. I'd literally go around to every single other engineer in the building and ask if they needed help. If they didn't, I sat at my desk staring at the guy in the cubicle next to me. But I couldn't leave. Did I mention it was boring? I used to fantasize about what it would be like to be a teacher, and I knew that if I could make it as an engineer, really how much harder could teaching be.

Then I became a teacher. I will forever say that teaching has to be one of the hardest jobs in the world. Not only do teachers supervise thirty to forty kids per hour, but they also are charged with bringing these kids up to proficiency in the academic subject matter in which they're experts, while simultaneously acting as a counselor, a coach, and a parent. But people who have never been a teacher don't know, just like people who have never been a parent don't know the ups and downs of parenthood. I know I don't need to tell you how hard your job is, but I feel compelled to tell you ways to maintain your sanity and take care of yourself throughout the school year.

During my first year of teaching, one of my colleagues shared a version of this image with me:

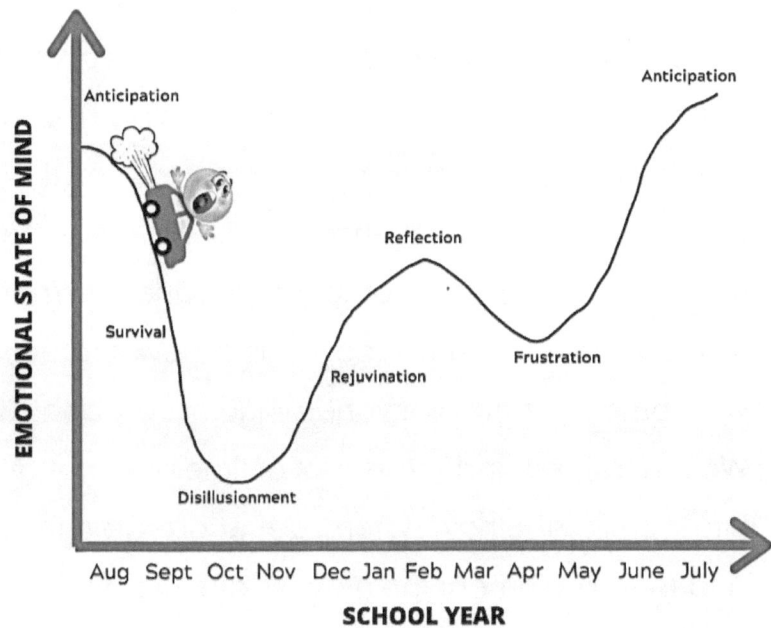

It's supposed to illustrate the emotional state of first year teachers. After having lived through thirteen school years, I can definitively confirm this represents *first* year teachers' states of mind, as well as *all* teachers' states of mind throughout the school year. The job is hard. Really hard.

We've talked a lot about the myriad challenges teachers face. Another challenge, which could be a whole separate book on its own, is instructional strategies. Over a thirteen-year period as a teacher, vice principal, and administrator, I've taken quite an extensive list of education courses, workshops and seminars — the amount equates to complete madness. What tends to happen in education is a new strategy will come up and everyone will jump on board. We get trained and invest a lot of time and energy into learning the new system, get acclimated to it, and then implement the new system. We do all of this only to have it replaced the very next year with something else.

The Cycle Starts All over Again

That's just the instructional strategies or how to deal with the academic content. You can't even *get* to the academic content until the behavior is under control. It's interesting how there are *so many* instructional strategies and trainings out there, yet virtually nothing on dealing with student behavior. So, if you can manage to get a handle on the latest and greatest instructional strategy, you'll be lucky if you can utilize it in your classroom because there is literally a social or emotional crisis happening with at least one student in each class every day. We all know that it only takes one student to change the entire dynamic of your classroom. So, we're devoting this chapter to taking care of yourself. Ever notice how on airplanes, if the sh@# is going down, you're instructed to put your own oxygen mask on before anyone else's? We need to take that same approach in education. So, how exactly do we do that?

Acknowledgement of the Struggles

Have I mentioned yet that this job is hard? Recent research in neuroscience has shown that each of us has mirror neurons in our brains. A perfect example to illustrate this is yawning. How many times have you seen a person yawn, then you feel compelled to yawn shortly thereafter? The culprit? Mirror neurons. What this means for us is that as human beings, if we notice someone else in the world is happy, it's contagious and has the ability to make us happy. The flipside is true as well. If we are surrounded by people who are sad or angry or suffering from anxiety or depression or hormonal teenage overload, our mirror neurons have the ability to absorb and mirror what we're surrounded with.

Additionally, new research has shown that people who work in "helping professions," like nurses and teachers, often suffer from what's called *secondary trauma*. The three different types of secondary trauma are compassion fatigue, vicarious trauma, and job burnout. This is a very real thing. If you don't take care of yourself, you won't last in this profession. Take time and acknowledge the reality of your job. Once you do that, take care of yourself.

Know Your Triggers

In Chapter 7, you went through an activity to identify what your triggers are. Go back and revisit that section so that you can begin to better identify your triggers. You have a life outside of work, and yet the expectation is that you show up to work and maintain a level of professionalism, leaving everything you might be going through in your personal life at the door. It's time to start asking the question: is that realistic? Yes and no. It is imperative to check your emotions before you show up to work, and if you're feeling emotions right under the surface, take a step back. How does "taking a step back look?" Worksheets, informational content-related films, etc., and at the very most, utilizing your sick days.

If you're feeling adventurous and want to take it one step further, lead a discussion with your students about what triggers are (sights, smells, tastes, sounds, feelings that remind you of a negative or challenging time or situation in life and cause your emotions to escalate quickly and unexpectedly). You can make this part of your fifteen minutes of non-academic time so that when you are triggered in the future, you can simply say to your students, "Sorry guys, I'm being triggered. I need a minute." I did this a few times with my students, and you wouldn't believe how compassionate they are when you tell them this. Again, it's the human connection piece. They see that you aren't perfect and that you too get affected negatively by life. That's when their empathy and compassion kicks in. They'll let you take a breather, and everyone will be better off for it.

Using Worksheets and Academic Films

The 2012-13 school year was a rough one for me. I was still working at the very high-performing charter school with kids who could learn from anybody — but the year started with our department chair suddenly passing away. This teacher was so loved and he passed suddenly and without warning. He taught Geometry and I taught Algebra 2, so that year I had all his students from the prior school year. When we found out that he passed, it was up to us provide the support to students when the announcement was made. It was hard. Really hard. Kids were crying and in shock, and all I could do was pace back and forth in my classroom, walking around to give hugs to anyone who was visibly upset. When the announcement was made, I stopped math instruction. All together. There was absolutely no way I was going to expect that these kids were going to learn solving quadratic functions right after hearing that one of their favorite teachers just passed away.

I used their reactions and body language and energy the following few days to decide what activities to provide. It was at least a few days of very simple review math worksheets, and eventually once the sting was gone, we got back into the content (and still managed to be at the higher proficiency rate).

Then about a month after my department chair's passing, I went through a separation and eventual divorce from my husband of ten years. I got the worksheets back out again and relied on Schoology and YouTube to help with the instruction. Am I proud of using so many easy review worksheets during this particular year? No. But had I tried to carry on as if nothing happened, I could've boiled over and had a much worse reaction and outcome. Remember, you're only human. It's not going to kill the kids to have worksheets a few more days than normal in one school year. Take care of yourself.

Using Sick Days

When all else fails, take a sick day. They are there for a reason. No, you do not have to be physically ill. You can take a mental health day and still use a sick day for it. Remember the situation where I told two students to "shut the f@#$ up" and "grow the f@#$ up?" I took a sick day following that incident because I realized just how much I neglected myself during that stretch of the school year and was mentally frazzled.

When in Doubt, Walk it Out

We've already talked about how kids naturally push your buttons and that it's part of their job description. But if you're taking care of yourself with various proactive measures, you'll be much better equipped to handle the confrontations that arise daily with your students.

Even with all the self-care in the world, however, you'll inevitably be faced with a situation in your classroom where behavior is escalating, and you need tools to be able to deal with it while keeping your job. So how do you do that? Remember this expression: When in doubt, walk it out.

That doesn't mean just leave your classroom and walk away, leaving campus never to return. It means say to your students, "Hey guys, give me a minute, I need a break." Step outside for a few minutes with the door cracked open so they are still in your line of sight. Take a few deep breaths and attempt to gain your composure. If after a few minutes you're unable to, simply go back in your classroom and call one of your colleagues who may be on prep or the office and say you need someone to watch your students for a bit. Once a colleague shows up, take a quick walk around campus or go to the bathroom to compose yourself before returning to your classroom.

Practice Proactive Self-Care

Okay, so we've spent some time talking about reactive ways of taking care of yourself in the heat of the moment when working in your classroom and not feeling well. The way to prevent reactivity as much as possible is to be proactive. I don't need to tell you that there are many ways of being proactive in terms of self-care. But what I am going to ask you to do is look at the list of proactive self-care options below and circle the options that are most appealing to you.

- Sleeping
- Enjoying the sunshine
- Cooking
- Writing

- Drawing
- Talking to yourself
- Cuddling with pets
- Taking a walk
- Taking a bike ride
- Cleaning
- Reading
- Gardening
- Getting a hug
- Talking to people
- Watching a movie
- Listening to music
- Telling jokes
- Trying a new craft
- Having a dance party
- Swimming
- Practicing yoga
- Making a gratitude list
- Practicing mindfulness
- Practicing positive self-talk
- Spending time with friends
- Golfing
- Taking a bath

- Meditating
- Finding shapes in clouds
- Hiking
- Jogging
- Lifting weights
- Practicing mindful breathing
- Creating mandalas
- Playing games
- Enjoying a meal with someone you love
- Having a BBQ
- Burning your favorite candles
- Punching pillows
- Getting a manicure or a pedicure
- Getting a massage
- Painting
- Celebrating small successes
- Stretching
- Talking to a counselor

Once you've circled the options that are a best fit for you, make a commitment to practice one self-care activity each day for the next month. I can tell you that for me, walking, yoga, meditation, celebrating small successes, and spending time with my family are at the top of my list and things that I practice every day. The way you can really make this stick is by scheduling it into your calendar. I have a block of time scheduled each day to practice yoga, meditation and exercise. When people call to set up appointments or book sessions or trainings with me, I am in a "meeting" during the blocked-out self-care time. It's a meeting with myself, but again, if you don't take care of yourself, there's no way you will be able to take care of anyone else. As a teacher, you are one of the most important people in the world. Treat yourself that way!

Have a specific empower-yourself question, comment or concern? I want to know what you think!

Head to **withheartproject.com/contact** so we can chat!

Chapter 10: Relationship-Build

As Kaylie angrily shuffled out of my classroom she took the sign-out clipboard by the door and threw it toward me and yelled, "You f@#$ing bitch!" The clipboard hit the front wall and snapped in half as it made its way to the floor. The entire class stared at me, waiting to see what I was going to do next. I was surprisingly calm. This was my third year of teaching and I had already instilled many of the behavior management techniques outlined in Chapter 4. I checked in to make sure no one got hit by the flying clipboard and carried on with the lesson. Once class was over, I made my way down to the vice principal's office to check in with Kaylie.

She was sitting at a table outside of the vice principal's office, and I sat right next to her and calmly said, "Okay, what's going on?" She was silent. The incident that caused me to send her to the office in the first place was disruption during the lesson. She was talking, I asked her to be quiet, she said she would but ultimately kept disrupting class and the lesson. After the third attempt to get her to stop talking, I followed my own protocol and sent her out, out of respect for the students in the class paying attention and trying to learn. It was a minor incident with a *major* blow-up.

I said to Kaylie, "Listen, I'm sorry that I had to send you out of the classroom. Do you understand why I had to send you out?" She started to cry. I continued to probe to see what was going on under the surface.

"Kaylie, what is going on?" I asked.

"Nothing," Kaylie said.

"I don't buy that. What are the tears for?" I asked.

"My mom said that if I got in any more trouble at school that she would send me back to juvie," Kaylie said shamefully.

I felt terrible. I had a great relationship with Kaylie. She was the student who had her mom come to school year after year so that the counselors would rearrange her entire schedule just so she could be in *my* math class. She wouldn't stand for any other math teachers. This was the third year in a row that I had her, so when she talked and behaved disrespectfully during the lesson, it was confusing to me. She was typically a student I would turn to and tell other kids to behave like her. But this day, she was on a roll and pushing my buttons.

"Oh honey, I'm sorry," I replied. "What was making you talk the way you were? You're typically one of the best students in class. It was frustrating to see you behave in the way you were. What was going on?"

She kept her head down and wouldn't look at me, "I don't know. Mike was saying something funny, so we just kept talking."

I said, "I understand," then paused. "Do you understand why I had to send you out of class?"

She nodded and continued crying.

I gently asked, "What do you think your mom will say when I call her to tell her about this?"

She quietly replied, "She'll be mad and then send me back to juvie. All she ever says is that I'm never going to graduate from high school, so I may as well go back to juvie, so I have a safe place to stay."

I was shocked and disheartened, but replied, "Well, I think you're going to graduate from high school. You are incredibly bright and an all-around wonderful young woman. Do you think it would help if I told your mom about the situation but then explain that we got it handled and there doesn't need to be any further punishment or consequence?"

She looked defeated, "Maybe."

So, I called her mom and explained the situation, then asked that her mom not instill any other consequences. Her mom was understanding, and Kaylie returned to class the following day.

You might be thinking, "What the heck are you doing Kristen? That doesn't sound like consequences or punishment or behavior management. You're crazy!" But here's the thing. When Kaylie returned to class, she was the model student she normally was. There are many things about this interaction that I'd like to explain further, starting with upholding same expectations for *all* students, including your best students.

Good Student, Same Consequence

As I mentioned, Kaylie was a model student in every sense of the word, at least she was in my class. Many teachers give special treatment to their best students simply because they aren't the typical troublemakers. If a student is getting an "A" in your class, why should they be punished and/or given the same consequences as the kid who sleeps, talks through lessons, and has an "F" in your class? One word: equity.

I only had three "class rules" posted on my wall in my classroom, and they looked like this:

1. No eating, drinking or gum chewing allowed.
2. No electronics of any kind.
3. Respect yourselves, your peers, the teacher and the classroom.

When it came right down to it, I only truly upheld No. 3: Respect. I actually used this "class rule" list to build relationships. During the first week of school, I would always go over the rules, and just let them find their way. If kids were eating in class, if it wasn't disruptive or destroying my classroom, I tended to turn a blind eye; same with electronics. Because of this, kids viewed me as a "cool teacher," a "lenient teacher." I may have had that reputation for those two rules, but one where I was definitely not cool or lenient was the respect rule. So anytime *anyone* began disrespecting anyone or anything in the classroom space, consequences would be enforced and upheld, regardless of who was being disrespectful. Why? Because as human beings, we all deserve respect.

Our history is rife with examples of disrespecting people because of something as minute as the color of their skin. This is 100 percent unacceptable. In the same way that this type of discrimination and inequity is unacceptable, so too is allowing a straight "A" student to get away with being disrespectful, bullying, bringing drugs or alcohol to school, or any other type of infraction worthy of consequence.

It was *this* mentality that allowed me to build and foster relationships with *all* of my students over the thirteen years I was in either the classroom or and office on one specific school site. Another example is a student with whom I had built a strong relationship during my time as a vice principal. She was struggling with sexuality issues and questioning whether she was homosexual. She found safety and solace in my office because I accepted her and helped her navigate these issues. Then one day she did something stupid. She brought some sort of pills to school and passed them to her friends in plain sight. I did a full investigation and confiscated the drugs and ended up suspending her from school for this action.

Unfortunately, I have seen teachers and administrators make special accommodations and/or turn a blind eye to things like this simply because they liked the kid. Well, I loved this kid, Stacy, to death, but I still upheld the consequence as outlined in Education Code 48900(c) (Unlawful Possession of a Controlled Substance). But here's the question I have gotten time and time again from educators: How do you build relationships while still toeing the line?

Building Relationships While Toeing the Line

One of the most effective ways of building and fostering relationships with your students while *still* upholding rules and consequences comes down to *the way you communicate with your students*. As an administrator, your job is to do a complete investigation to see exactly what happened, get witness statements from any and everyone who was in the vicinity and/or directly involved, and then let the truth and evidence guide the consequence and outcome. So with the previously-discussed student who brought drugs to school, that's what I did. Not only did I see the drugs in her hand with my own eyes, but every witness statement I took from her friends implicated her guilt. So, when it was time to tell her exactly what the consequence was, I did so with empathy. This word is *so important* that I'd like to write it again, on its own line.

Empathy. Here's how the conversation went after I completed the investigation and determined that she did, in fact, violate Education Code 48900(c).

I brought Stacy into my office and asked her to sit down. I knew this was going to be a difficult conversation, but one that I had to have, given all the evidence I collected. She was visibly nervous. I said, "Okay, Stacy, so here's the situation." I pulled out the District Student/Parent Handbook, which had a section on code of conduct and consequences for a variety of infractions.

I continued, "So, you know that we've been doing a lot of looking into the situation with you bringing the drugs to school."

She nodded.

I said, "Unfortunately, Stacy, I have to suspend you from school for this. I am really sorry."

She began to cry.

I opened the Student/Parent Handbook to the page about being in possession of a controlled substance and explained how the chart worked. This chart outlined what consequence would be given for various infractions. I ran my finger down the page and showed her the section on possession of a controlled substance and said, "This is how this works. This chart is like the 'laws of school.' In it, it shows what type of consequence is in place for different rules that are broken. When you brought this substance to school, you were what is called 'in possession of a controlled substance,' and the consequence for this is suspension."

She sat there and stared at the ground continuing to cry.

I asked, "Do you have any questions?" She nodded 'no' and continued to cry while I explained, "Honey, I don't have a choice. This is the rule and I *have to* suspend you even if I don't want to. I am so, so sorry."

We sat in silence for a few moments, and she anxiously asked, "Do you have to call my mom?"

I replied, "Yes, unfortunately, I do have to call your mom."

She cried even harder. I asked her if there was a specific way she'd like me to break the news to her mom. We came up with a way of including mom and dad that was the least hurtful and difficult to her, then called her mom and continued with the process. I explained in as much detail as I could what was going to happen and did so with as much empathy as I could.

Now you might be thinking, "Empathy? I've never brought drugs to school; how can I have empathy for that?" Well I never took drugs to school either, but I *do* remember what getting in trouble felt like, and what disappointment from my parents looked and felt like, and it was not fun. So, when I say that you need to deliver consequences with empathy, what I mean is, tap into how you felt when you got in trouble in your youth and really remember and **feel it**. For me, I can go back in time and imagine a situation where I did something stupid and my mom was *so mad* that she yelled and me and wouldn't even look at me or talk to me for a few days. It *hurt*. So anytime I have to issue some sort of consequence to my students, I do so with empathy. I think about how I would like to be treated in times where I made mistakes and proceed with compassion and sensitivity. It is *this* that allows me to foster wonderful relationships with my students while still toeing the line.

The proof of this strategy comes with time. If I rewind to the situation with Kaylie, she was pissed at me for sending her out of class, and she was angry for a while. But she eventually got over it and realized that my enforcing consequences with her was an act of caring. My relationship with her was so impactful that after I moved on to a different school, she ended up showing up at that school just so she could say hi and check in with me. She also wrote me a three-page letter thanking me for everything I did for her. I didn't get confirmation that I was on the right track until years after doling out consequences that caused anger, hurt, sadness, and resentment.

Same with Stacy. She refused to talk to me for a long time. It may have been a month or so before I was able to really build that bridge back. I had to talk with her during that time for various situations that came up on campus, and during the month that she wasn't talking to me, our conversations were short and curt. However, every time we did have a conversation, I reiterated to her that it was okay that she was mad at me, and it was expected, and that I was there and ready and waiting if and when she needed to lean on me again for support. She eventually did come back and lean on me for support, and our relationship got back to normal again.

The thing is, when you enforce consequences and expectations, students know deep down that it's coming from a place of caring for them and their well-being *if* it's done with empathy and compassion. There's a huge difference between a teacher who kicks kids out of class and never follows up, and a teacher who kicks kids out of class but takes the time to have a follow-up conversation with the student afterwards. Of course, this whole process is easier (or easiest) once you've built that relationship with your students in the first place. So how do you do that?

Be Vulnerable (Some More)

The idea of being vulnerable in front of your students is scary. Really scary. But it's also really crucial. We talked in previous chapters about how we are dealing with *people*, and people are messy. These aren't little robots who are just there to "give you" high test scores. They are *people*, and they're experiencing all kinds of feelings and emotions, and life situations trigger these feelings and emotions under the surface. If you can find any opportunity to weave vulnerability into your classroom, not only will you build incredible relationships with your students, but you'll also create a safe space for your students and an environment where *learning* can take place.

Tell Stories

Allow your students to get to know you as a person. Show them pictures of your family and weave your own life stories into your lessons every chance you get. I did this when teaching the addition and subtraction of integers through a creation I called the "Mindful Human Number Line."

On this number line, zero represented a "neutral" mood, +10 represented a "happy" mood, and -10 represented a "bad" mood. So, I'd use my own stories to illustrate how adding and subtracting integers worked in life. An example was, "I was going along one day and was at a seven. Overall, a pretty good day. Then my cat got sick and we had to put her to sleep, which was the equivalent of ten bad things happening. So, if my day started at a seven, and the bad things happened, where does that put me on my number line?" When putting academic concepts into life stories (and yes, your own life stories are best, so kids get to know you, the better it is for everyone), not only are you building that relationship with your students, but you're teaching academic concepts in a way that is understandable.

I know what you're thinking, "Do we really need to use this touchy-feely bull@&#% to build relationships with kids?" **100 percent, yes.**

Whether or not you want to admit it, the one thing we know about humans is that we are hardwired to connect with each other. The bottom line is you *have to* teach academic content, and we've already talked about how academic content can be learned via YouTube. The one thing that sets you apart from any and all teachers on YouTube, is your ability to connect with them. How do we connect? Feelings, emotions, life stories, good/happy life situations, and bad/sad life situations. Kids are like sponges and soak up everything you do and say. Why not let them soak up your life stories and lessons while also teaching academic content?

"But I Don't Want Them to Walk All Over Me"

I get it. I totally understand how scary it can feel to "let them in," let them see you as a person, as an actual human being. We see students destroy each other daily for things like their clothing choice or the way their feet look or the words they use. How can we possibly think that if we open up about our personal lives and experiences, that they won't do the same to us? But that's just it. They won't. The reason they lay into each other and are so cruel to each other is because they don't have examples of *how to* interact with each other in a respectful way. If you model that for them, my gosh, you are doing wonders for them and for yourself!

It seems counterintuitive, but it works, I promise. Kids want to be praised, recognized, and encouraged. Have you ever heard the expression, "Negative attention is better than no attention?" Right now, kids are craving attention and notoriety anywhere, including in your classroom. Many kids go home and have absent or neglectful parents. They're craving recognition and attention and guidance as to how to be people in this crazy world. They will continue to act crazy and disrupt your class if you continue to pretend that they aren't human beings and that your subject matter and academic content is more important than them as human beings. Here are a few guidelines for building relationships while still toeing the line:

- **Care about them as *people* first, scholars second.** Every single day when kids come into your classroom, you should be scanning the room, reading the energy of the room, looking for overall moods, happiness levels, sadness levels, withdrawal in students, etc. If a kid looks distraught, *ask them* about it. It doesn't have to be right in the moment you notice them being distraught (you do have a class to teach after all). When you get to a point where the rest of the class is occupied, pull them outside and check in with them. Remember those two questions ("Are you okay?" and "What's going on?").

- **Ask questions**. During a training, I had a teacher ask me, "Aren't you afraid you're being nosy?" My response, "Nope." There was an experiment done a couple years back where an individual followed one specific middle school student around to his classes all day long. What this person found was astounding: this student went through his entire day without one single person talking to him. Keep that in mind every time you think you're being nosy. What's "nosy" to you may be a connection and much-needed lifeline to the student.

- **Enforce consequences across the board, equally to everyone**. If you don't allow food in your classroom, don't allow food for everyone. Not just the "bad" kids.

- **Pick your battles regarding class rules**. Think about this one mindfully. I worked with a teacher who required that all her middle school students push their chairs in every day. If they didn't push their chairs in, they got a detention. Let's be honest. Do you really want to spend your time and energy being a chair police? I'm guessing not. What kind of fun is that? Not only is it exhausting to make sure 30 to 35 kids every hour push their chairs in, but then for those who don't, now you have a mountain of paperwork to fill out. Everyone loses in this scenario. When it comes right down to it, the one rule that really matters is respect! Respect, empathy, and compassion. If there are any rules truly worth upholding and enforcing consequences for, it's this one. Treat each other with respect, practice empathy, and compassion.

Not only does this rule make sense, but it helps foster amazing relationships and culture in your classroom. I had a student say to me one year, "Miss Miller, the only place at this school that I feel safe is in your classroom. Everywhere else I don't feel safe." While I'm honored that she felt that way, it made me incredibly sad to know that the rest of the school was like a mine field she had to survive on a daily basis.

- **Foster relationships with your students' parents**, especially those of your most challenging students. Pick out one student who will likely be challenging (the kid who tests boundaries at the *very beginning*) and get to know his/her parent well. Make a phone call to this one parent within the first week of school and introduce yourself. Let the parent know you're excited the student is in your class. Tell them you'd love to know more about that student's interests, hobbies, and goals for the future. Make it a point to find something positive about this student and communicate it to the parent as soon as you see it happening. I would often ask my parents if I could send them texts instead of calling for small things, as it was easier given the nature of my busy job, classroom and student privacy and confidentiality. If you're asking when telling them something positive about their kid, 99 percent of the time they'll say, "Yes, of course!" The reason this is so important is because inevitably, this challenging student will continue to be challenging throughout the year. If you take the time to recognize the positive not only to the student but also to their parent, you'll be building up a relationship that you can lean on when times get tough. The *only* reason Kelly's mom didn't report me for telling her kid to "shut the f@#$ up," and "grow the f@#$ up," was because she knew I had Kelly's best interest at heart. How did she know that? Because I would text her a compliment about Kelly once a week or every other week. I wasn't only contacting her when Kelly was being challenging.

We've gone through a lot of ways to build relationships. I know the idea of this is completely scary and foreign, and you're not sure you'll be able to do it. But you will, and I'm here to support you. Please don't hesitate to reach out anytime for guidance or support. You can find me at www.WithHeartProject.com. Now, go build some relationships!

 Have a specific relationship-building question, comment or concern? I want to know what you think!

Head to **withheartproject.com/contact** so we can chat!

Chapter 11: So, Now What?

According to *The Wall Street Journal*, "Teachers and other public education employees, such as community-college faculty, school psychologists and janitors, are quitting their jobs at the fastest rate on record, government data shows." Why are they leaving? They are citing things like:

- Tight budgets
- Small raises
- Poor working conditions
- Student behavior
- Parent behavior
- Politics
- Lack of support

So, now what? Are you going to be just another statistic or are you ready to take massive action and make a massive change? It all comes down to how much discomfort you're willing to live with and live through. I know you became a teacher because you had a deep love of and passion for helping kids. I know that love and passion is still in there deep down, probably buried beneath all the shifting expectations, bad student behavior, bad parent behavior, and lack of money. So, you're at a crossroads. You can:

- Continue in the teaching profession like you've been this whole time, disgruntled, frustrated, feeling alone, and not making the changes you need to make to better meet your needs and your students' needs.
- Continue in the teaching profession with the willingness and bravery to shift your teaching practice.
- Leave the teaching profession.

I'd like to explore each of these options a little bit further. In this book, we've gone through a process that will help get you to a place where you are your best teacher self. I am 100 percent confident you can apply the principles in this book and make massive shifts to your practice all by yourself. I truly hope that you will do so.

I'm also 100 percent confident you *will* make massive shifts to your teaching practice with my continued help, guidance, and support. Let's unpack this even further. Oftentimes the reasons people don't make massive changes is because they don't have the internal confidence to actually make those changes. They keep telling themselves things like, "Oh, you're really not that unhappy," or "This profession is so messed up that my making changes won't make a difference," or "Everyone else is still doing it the old, meaningless way, so I will too." All these sentiments have a tinge of hopelessness and helplessness. I know that deep down that's not you. You got into this profession because you love kids, because you want to inspire and empower them. But the system is standing in the way of you doing that. So, you can run away from it and keep doing the same old thing, or you can take a risk, and go on this journey with me to revolutionize education.

It's not going to be easy. By attempting to take massive action and make massive change, you're likely to run into many obstacles, including:

- Not *actually* shifting your practice to better meet you and your students' needs because it's too hard.
- Not *actually* implementing new behavior management strategies because it's too hard.
- Not attempting to open the lines of communication with your administrators because it's too hard.
- Not being able to handle political pressures, and thus leaving the profession either literally or figuratively by becoming complacent and reverting to "old or set" ways.
- Not actually looking at yourself and attempting to identify your own hang-ups and barriers.
- Not actually trying anything new because it's too scary or too hard.
- Not actually taking care of yourself because you don't "have time."

- Not taking the time to build relationships, ease off the curriculum, because you don't think it'll help.

All of these seem like compelling reasons to stop right here and run the other direction. But I see the fire inside you. I see how much you care and how you want to make these changes. So, let's look at what could happen if you take a risk and start making massive changes.

You could be happy at work and not wake up each morning dreading going to your job. You could be healthier overall, improving relationships both within and outside of your classroom. I can tell you that when I was going through some of the challenges I outline in this book, it took a huge toll on my marriage. My husband was able to do the "cheerleader" thing to a point, then he couldn't handle my frustrations and complaints anymore. When you implement these practices in the classroom, you'll finally be able to take some time to have *fun* with your students, get to know your students, not be so stressed out that you can't imagine a single second wasted on non-academic instructional time. And my data proves that *less* academic instruction is better for your students and will lead to higher academic gains than *more* academic instruction. So, there you go! I've let you off the hook. Go and enjoy getting to know the people sitting in your classroom and don't feel guilty about it.

You could be part of this much-needed movement in education, and a critical part at that. One by one, if we all start shifting our perspective and teaching practice to better meet our needs and those of our students, we will be able to shift this entire education system so it starts meeting everyone's needs, including the needs of society (that's an entirely different book topic).

By taking this risk and massively shifting your practice, you are providing a new model for education. You'll be a huge role model for how to deconstruct a 100+ year-old system and rebuild it for the twenty-first century. How inspiring is that? If, however, you're not inspired, here's what happens if you don't contribute to solving this problem:

- Your constant anxiety and depression stick around.
- Your marriage or relationship continues to suffer.
- Your home life with your kids will be at jeopardy.
- You may have to change careers.
- You may have to go back to school or get additional training for a new career (think of the costs associated with that).
- When shifting to a new career, you'll likely start at the bottom of the pay scale, losing all the income you've built up.
- You'll continue to educate generations of kids who will never reach their full potential.
- You'll make me sad when I *know* you can do this and will be *so much happier* on the other side. So, this is

the moment of truth. Are you in or are you out? Are you ready to move mountains with me?

Chapter 12: Conclusion

So here we are. We've been on quite a journey together. We've laughed together (hopefully), cried together (maybe), and are now looking forward to what lies ahead. Before we do that, let's do a quick recap of everything we've gone through together. We learned The EMPOWER method and you've been given a variety of tools to empower you to be the best teacher you can be. Here's a quick review of what those steps look like:

E–Eliminate Bad Behavior

This chapter contained a variety of tools to help eliminate bad behavior immediately. We began with the ever-so-important *Love and Logic* strategies, which was all about choices and not going on the emotional roller coaster with students. Then we covered the importance of utilizing Positive Behavioral Interventions and Supports, much of which detailed, nitty-gritty examples you can find on <u>www.PBISWorld.com</u>, but the main takeaway was flooding your students with positivity, recognition, encouragement, and positive support. Next, we talked about Restorative Practices, and specifically the use of affective language and conflict resolution. Affective language is essentially talking to kids in a way that allows them to see how their actions and/or words *affect* you and/or other students. Conflict resolution is done best through a non-judgmental series of questions to all parties involved and can help minimize conflicts in your classroom. The component of Trauma-Informed Practices we discussed include teaching kids about their brains and what is happening during conflicting situations with students (fight or flight), as well as specific strategies for various triggering situations that may arise in your class. General communication skills were covered to help you realize the importance of things like facial expressions, tone, and body language. We learned that the combination of all of these "buzz word" strategies make up the overarching umbrella of Social-Emotional

Learning, and that utilizing these strategies can be the difference between successful students and unsuccessful students.

M—Maintain Positive Administrative Relationships

In this chapter, we covered what the vice principal job looks like from the teacher's perspective and I shared the first of two traumatic experiences that shaped my career: a student slicing off part of his finger in my class. We talked about what a day in the life of a vice principal could look like and shed some light on the pressures and politics they face in various behavior-related disciplinary procedures. We analyzed how what appears as lack of teacher support can sometimes be much deeper than that, such as threatened or pending lawsuits where site-level administration has zero control. We also went through a series of steps to help teachers learn to take a step back and recognize that administrators are people too, just doing the best they can. And finally, we reminded administrators that they also need to take a step back and recognize that teachers are people too, doing the best they can, and that ultimately, we're all on the same team: team student success!

P–Push Through the Politics

This chapter discussed series of situations that are highly politically-charged and largely out of our control. I described in painful detail the second traumatic situation that drastically shaped my career path: being fired for literally no reason. We walked through the importance of maintaining dignity and grace when facing political pressures. And finally, you were given a set of steps to help *you* maintain dignity and grace when facing political pressures.

O–Own It

This chapter was devoted to self-exploration. It's very easy to swim along in the teaching profession and not make any changes to your teaching style. In this chapter, we looked at our own behavior management style and how the various behavior management styles can play out in your classroom. We also looked at implicit bias, which is the unconscious attitudes and stereotypes that affect our understanding, actions and decisions. You were encouraged to take the implicit bias test and share your results. We also went through an activity where you were to identify your own triggers, and then we went through typical ways you respond to these triggers and more constructive ways to respond to these triggers. If you can't own your hang-ups, biases and management styles, you'll never be able to try something new.

W–Why Not Try Something New?

In this chapter, we visited another impactful situation that led me to the discovery that "less is more" when it comes to academic instruction time. We talked about the pressure to cram every minute from bell to bell with academic instruction, and that this very pressure causes our students to learn less. I introduced you to the Wellness

Wednesday idea and only teaching academic content a *maximum* of 75 percent of the time. Then we discussed various ways to fill the remaining 25 percent of instructional time by:

- Community-building circles
- Mindful breathing
- Mindful walking
- Mindful listening
- Wellness writing
- Daily greetings
- Taking time to learn your students' names
- Celebrating your students
- Snowball fights
- Sharing your own stories
- Being vulnerable

E–Empower Yourself

In this chapter, we stressed the importance of self-care and acknowledged that the teaching profession is *hard*. **Really hard.** So hard that I put it as the second most difficult job in the world. So, in this chapter we talked about the importance of acknowledging the struggles of the profession, knowing your triggers, using worksheets and/or academic films when you're not "feeling it," using sick days, walking it out during heated and escalating classroom interactions, and challenging ourselves to take better care of ourselves.

R–Relationship Build

The final step in the EMPOWER process is all about building relationships. In a world of technology everywhere, this is *critical* to being successful long-term as a teacher. I shared my experience with one of my students who I doled out consequences to, but also had a great relationship. We talked about the importance of equity when it comes to discipline and consequences. We discussed specific ways to build relationships with students while toeing the line (Hint: using empathy and compassion while giving consequences). We talked again about the importance of being vulnerable and how to use your own life stories embedded in academic content to help build relationships, and we covered how practicing these methods will not make your students walk all over you. In fact, it will do quite the opposite: It will make them respect you and work harder for you than they ever have before.

This book was jam-packed with specific strategies to help you make more out of your teaching practice. My wish for you is that you start with yourself, and ask yourself two questions:

- *What is holding me back?*
- *What fears do I have that are keeping me from being my best teacher self?*

Take a minute and imagine a school system where the vast majority of students want to be there, in your classroom (in all their classrooms), come prepared to learn, and work as hard as they possibly can to please you and be a good student. I can definitively tell you that this is not a dream. This is a reality. This was my reality for the thirteen years I was blessed to work with adolescents. I came to the realization that I loved education and teaching so much and was having an incredible amount of success with my students that I wanted to share these strategies with other teachers just like you, so you could experience the same level of success and joy and passion the teaching profession has to offer. Together we can move mountains by EMPOWERing ourselves to be our best and meet our own needs, so we can better meet the needs of our students.

It all begins with you and me.

Right here.

Right now.

Let's do this!

Acknowledgments

First and foremost, thank you to **Nic** and **Sophie**. You laughed with me, cried with me, and managed my ridiculous reactions to the events as they happened throughout this my time in the world of public education. I appreciate you more than I can express in words.

Mom and **Dad** – I ended up here strictly because I had two amazing, supportive, wonderful parents to teach, guide, and mentor me through all of the ups and downs in life. I would not be here without your strong words of encouragement and guidance through all the trials and tribulations over my life. Thank you.

Steph and **Jessica** – I could not have asked for two more amazing, supportive, caring people to have as sisters. I love our relationship and feel so lucky to be able to lean on each of you. I love you.

Steve – Thank you for all of your support and guidance throughout many of the situations described in this book. You were such a rock for me for so many years – I am eternally grateful for you and your presence.

Joel and **Brian** – In different ways, you both have helped me so much both personally and professionally. I am so grateful for your support.

Dr. Angela Lauria and the entire **Author Incubator** team – I was wanting to write a book since my professional world changed so dramatically – thank you for providing the framework, guidance, encouragement and support to allow me to finally complete it. I am eternally grateful for you.

Ann – you and I worked together through a number of these years and leaned on each other during challenging times. I am so grateful for you as a colleague and a friend. Thank you for taking the time to read and provide feedback.

Jessie – you were and still continue to be one of my most memorable students. You were there during so many transitions and tragedies, sometimes I wonder how I would have survived those moments without your strength. Thank you.

Your thoughtful feedback has been much appreciated.

Terri – you were my right hand during my first few years of teaching. Thank you for your honest perspective and guidance back then, and continued perspective and support now. I appreciate you.

Jaime – there hasn't been a dull moment in our friendship – we met creating "dream" resources for math teachers and continue working together and evolving as educators. I am so appreciative of you and your friendship and support. Thank you for everything.

Michelle – thank you for your continued support of this work. You are an amazing educator and I am lucky to have you be a part of my world.

Donna – thank you for your continued partnership and words of encouragement and praise. I can't thank you enough for taking the time from your busy schedule to say some words to help guide future teachers. I appreciate you.

Amanda – thank you for taking a chance on me with this book – I am so grateful for your honest words and feedback. Let's band together moving forward and continue changing the teaching world!

My **fellow teachers** – I wrote this for you and continue to be inspired by each of you who show up and are present and want to do what's right for kids every day. You give me hope.

My **past**, **present**, and **future students** – keep using your voice, speaking up and standing up for what is right and helping us shape the way we do education for all those coming through the system. You matter more than you know.

About the Author

Following a three-year engineering career, Kristen Miller spent thirteen years in traditional education roles, teaching predominantly high school mathematics,  Advancement Via Individual Determination (AVID) and Career Technical Education (CTE), and serving as a middle school vice principal in Northern California. Seeing a huge need for high-quality social and emotional interventions, systems and supports, Miller created a youth empowerment organization, With Heart Project (WHP), to work alongside schools and districts promoting Social-Emotional Wellness (SEL), Restorative Practices (RP), and Positive Behavioral Interventions and Supports (PBIS).

In her inaugural year, she partnered with a high-poverty middle school in Northern California to create and implement practices, processes, and procedures to decrease student suspension rates and increase academic achievement. Her results were remarkable. Her efforts yielded a 79 percent growth in Common Core Mathematics achievement, a reduction in discipline and attendance infractions, an increase in GPA, and an increase in math and reading grade levels among at-risk students. She is currently a Doctoral of Education candidate in Transformational leadership. Kristen lives in Sacramento, California, with her husband, daughter and nine pets.

Thank You

Free Video Class: I have a companion series that goes with this book. You can head to www.WithHeartProject.com to sign up for it.

Free Coaching Session: Want to talk about your teaching practice and/or all things education and get my take? Awesome! Head to www.WithHeartProject.com and fill out the contact form.

Find Me Here:

Web: www.WithHeartProject.com

- ⓕ   @withheartproject
- ⓘ   @withheartproject
- ⓥ   @withheartproj